MOUNTAIN

The Earth series traces the historical significance and cultural history of natural phenomena. Written by experts who are passionate about their subject, titles in the series bring together science, art, literature, mythology, religion and popular culture, exploring and explaining the planet we inhabit in new and exciting ways.

Air Peter Adey
Cave Ralph Crane and Lisa Fletcher
Clouds Richard Hamblyn
Desert Roslynn D. Haynes
Earthquake Andrew Robinson
Fire Stephen J. Pyne
Flood John Withington
Gold Rebecca Zorach
and Michael W. Phillips Jr
Islands Stephen A. Royle
Lightning Derek M. Elsom
Meteorite Maria Golia
Moon Edgar Williams
Mountain Veronica della Dora
Silver Lindsay Shen
South Pole Elizabeth Leane
Storm John Withington
Tsunami Richard Hamblyn
Volcano James Hamilton
Water Veronica Strang
Waterfall Brian J. Hudson

Mountain

Veronica della Dora

REAKTION BOOKS

Alla mia mamma che ama le montagne

Published by Reaktion Books Ltd
Unit 32, Waterside
44–48 Wharf Road
London N1 7UX, UK
www.reaktionbooks.co.uk

First published 2016, reprinted 2017

Printed and bound in China by 1010 Printing International Ltd

A catalogue record for this book is available from the British Library

ISBN 978 1 78023 647 6

CONTENTS

Preface

Among all geographical features, mountains are commonly deemed to be the most permanent. Majestic and awe-inspiring, mountains are the first objects in a landscape to capture our attention. Their hard stone transcends human temporariness; it is an absolute mode of being. As such, mighty rocks and peaks are key protagonists of ancient cosmogonies and of most religious narratives.[1] In the West, mountains are also places that have helped to shape our perceptions of wilderness and of the sacred. Their cultural history is largely the history of our relationship with the natural and supernatural Other.

The wilderness and the sacred share two main similarities. First, they both evoke separation from the ordinary against which they are defined. Second, taken literally, they both cause bewilderment, wonder, displacement. Wilderness is psychological as much as it is geographical: it can be a state of mind and a state of the land. Yet for many today the term 'wilderness' evokes nostalgia, compassion, perhaps even a sense of moral guilt, rather than threat or mystery. In the Western geographical imagination, wilderness, like the sacred, is no longer boundless terra incognita, but, rather, a fragile archipelago to be safeguarded from the evils of modernity.

Wilderness and the sacred converge on mountain peaks, as do shifting attitudes to these two concepts. Historically mountains have repulsed and attracted. They have been appreciated and despised as sites of divine and diabolic sublimity, as abodes of gods and demons, of hermits and revolutionaries. Progressively

Aiguille du Midi.

Sunset on the Dolomites.

'domesticated', today they nevertheless continue to attract seekers of spiritual quietness and extreme emotions alike, as well as weekend excursionists seeking a break from the everyday. The reason is that mountains, like the sacred, are usually perceived as wholly other. 'Floating above the clouds, materializing out of the mist, mountains appear to belong to a world utterly different from the one we know.'[2] They demand separation from the ordinary, removal from earthly concerns – even for a brief moment.

The history of mountains is deeply interlaced with our cultural values, with our aesthetic tastes and scientific practices. Over the centuries mountaintops have been repeatedly lowered and elevated by the human imagination. New summits have been occasionally carved out of the land through scientific convention; others have been spoiled by intensive mining. Mountains, one might argue, are geological forms as much as they are social constructs. Yet there is something elemental about mountains that exceeds and transcends our attempts to ascribe meaning to them. How have conceptions of wilderness and the sacred been shaped through mountains? What do mountains teach us about the relationship between wilderness and the sacred? And, ultimately, what do they teach us about our relationship with our planet and with the transcendent? This book addresses these questions.

1 Mountain Matters

We have probably all been faced with 'a mountain of problems' at least once in our lives. Some of us might often find ourselves buried under 'a mountain of work'. And those with a bad temper might end up 'making a mountain out of a molehill'. Mountains punctuate our everyday speech as they punctuate the earth's surface. Whether physically or metaphorically, they strike us for their insistent materiality and for their prominence. In fact, the Latin term for mountain (*mons*) shares its root with the word 'prominence' (as do 'eminence' and 'imminent'). Likewise, 'amount' – a word that entered the English lexicon in the thirteenth century – comes from the Latin phrase *ad montem* (to the mountain).[1]

Mountains cover 24 per cent of the earth's surface and, while most of them pre-date human existence, their outlines have served as immutable backdrops to human history. Due to their prominence, some have provided different civilizations with distinctive landmarks or have been believed to be the seats of divinities (for example, Mount Olympus in ancient Greece, Mount Kailash in Tibet and the Navajos' Mount Taylor). Others have been used as defensive barriers from the Barbarian incursions into the Roman Empire to the Nazi Blitzkrieg of the Second World War. Mountain ranges like the Alps and the Pyrenees helped define the limits of emerging nation states in early modern Europe, whereas in the eighteenth century the Urals set the arbitrary continental border between European and Asiatic Russia.

Domenico di Michelino, *La Divina commedia di Dante*, 1465, fresco in the Cathedral of Santa Maria del Fiore, Florence.

Because of their sheer mass and fixity on the terrain, mountains continue to be conceptualized as natural barriers and to be metaphorically associated with obstacles and challenges. Dante imagined Purgatory as a mountain, and, during his lifelong journey, John Bunyan's Pilgrim likewise had to travel over a number of high places, including the Hill of Difficulty. In *The Atlas of Experience* (2000), a contemporary secular version of Bunyan's *Pilgrim's Progress*, the existence of a middle-aged person is dominated by the massive Mountains of Work and zones of 'high pressure', suggesting a prolonged series of daily challenges to be overcome through perseverance and endurance.[2]

'Destroy a country, but its mountains and rivers remain', a Japanese proverb says. 'If the mountain will not come to Muhammad, Muhammad must go to the mountain', we hear in different languages. The fact that mountains can hardly be moved is acknowledged in various cultures and is used to express persistence – of circumstances (and thus the necessity to adapt to them), of human habits and of belief. After all, we learn from the New Testament that only faith 'can move mountains' (Matthew 17:20 and 21:21).

Yet if mountains seem to be the most solid and stable of all geographical objects, semantically, they are probably also the

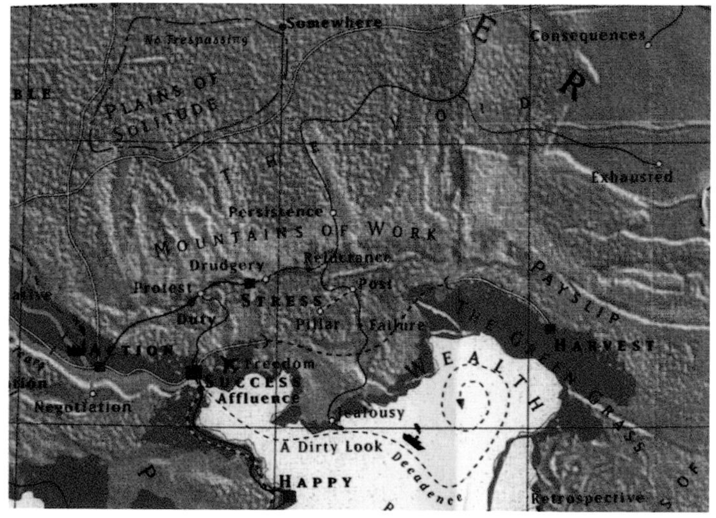

Detail of synoptic map of life from Louise van Swaaij, Jean Klare and David Winner, *The Atlas of Experience* (2000).

most unstable. It is very easy to point at a mountain, but it is very difficult to establish exactly what a mountain is. If you ask someone to draw a mountain, they will normally produce a well-defined triangular shape, but this is not what we find in nature most of the time. 'It is curious', John Ruskin observed, 'how rarely, even among the grandest ranges, an instance can be found of a mountain ascertainably peaked in the true sense of the word – pointed at the top, and sloping steeply on all sides'.[3] These archetypal mountains are usually those that attracted ancient myths and sacred narratives, as well as modern climbers. However, they are exceptions, rather than the rule.

The truth is that mountains come in all sorts of sizes and shapes: from the majestic grandeur of the Himalaya, standing at over 8,000 m, to the more modest yet no less spectacular Dolomites; from the sharp arrow-shaped summit of the Matterhorn to the pyramidal cone of Croagh Patrick; from the Andes, traversing the South American continent for 7,000 km, to the Apennines extending 1,200 km along the length of peninsular Italy. In the Aegean occasional isolated peaks stand out on the horizon, so that even modest altitudes have visual prominence and serve as steady landmarks for orientation, as they have been since antiquity.[4] In England and Wales, by contrast, mountains stagger in disordered sequences of lumps and bumps before they reach the valley floor.

We give the name mountain to snow-capped peaks and modest hills alike. The latter seem to be a recurrent urban feature around the world, from Montagne Sainte-Geneviève in Paris, rising a few metres higher than Notre-Dame, to Hong Kong's 'Peak' and Mont Royal, nowadays an urban park in the midst of Montreal, Canada. German city names ending in '–berg' (mountain) bear witness to this phenomenon.[5] We also refer to plateaus and sand dunes as mountains. The latter can reach up to 200 m and thus largely surpass elevations like Mount Alvernia, the highest point in the Bahamas (63 m), and Mont Cassel (156 m) in northern France. We give the name mountain to craggy structures of limestone or granite, as well as active and dormant volcanic cones, such as Mount Etna in Sicily and Mount Kilimanjaro in Tanzania, which, at 5,895 m, is the world's highest single mountain unconnected to a range. And, while we consider mountains the most visible of all geographical objects, we also number among them peaks we do not see, such as the submarine ranges crossing the abyssal plains of the Pacific – the least explored of all mountains. There are more than 10,000 submerged peaks that do not reach the sea surface, and these seamounts can rise up to 4,000 m, which is over 200 m higher than Mount Fuji in Japan and Mount Erebus in Antarctica and over 1,000 m higher than Mount Olympus.

Mountains are formed by the folding, faulting or upwarping of the earth's crust, caused, for example, by the encounter of tectonic plates or by the eruption of volcanic rock onto the surface. Great mountain ranges, such as the Himalaya, the Andes and the Alps, evolved by long and slow movements within the

Mount Kilimanjaro.

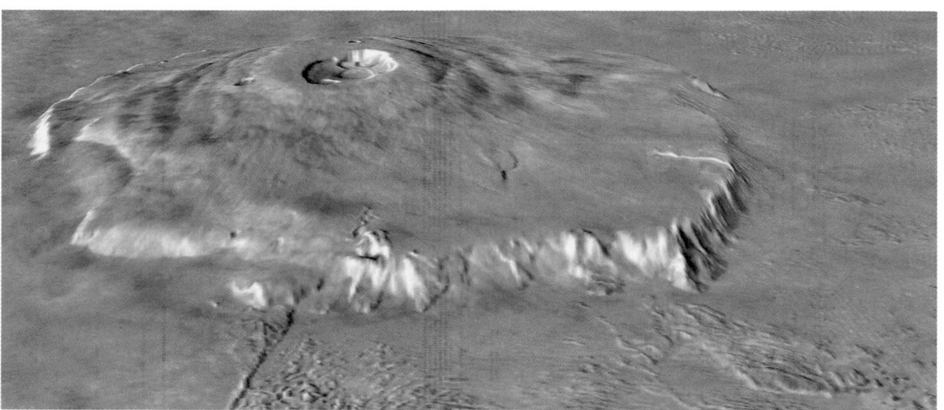

Olympus Mons, Mars.

crust. By contrast, volcanic cones such as Kilimanjaro or Fuji were gradually built up by layer upon layer of hardened lava and other erupted materials. Their different shapes and consistencies are thus determined by their underlying rock structure, as well as by different forces of erosion, such as frost and gravity. Mountains are therefore a characteristic feature and product of the terrestrial surface.

Yet we also label mountains as extraterrestrial heights – and, in fact, the highest summits are found on other planets, the tallest-known being Olympus Mons on Mars at 21,171 m, about three times the height of Mount Everest. Occasionally we call man-made piles mountains too. Some of the most intriguing examples loom on the horizon of the flat plains of eastern Estonia. Some of them are small pyramidal cones covered by pine trees; others look like miniature ranges. These artificial mountains are living mementoes of a bygone era. Resting on the emptied profundities of disused Soviet oil shale mines, they are the craft of female hands, for until the late 1960s it was the women's job to move the stones from the mines to the surface. Today these strange hybrid creations of nature and human labour have developed their own microhabitats and leisure industries – like most European mountains.

What is a mountain, then? Do we define mountains based on their altitude, on their size or on their shape? Are these criteria consistent across cultures? Have they remained constant

Oil-shale 'mountains' in Estonia.

over the centuries? Where does one mountain stop and the next begin?

Mountains are a relative category. While we can easily visualize a mountain, setting the boundaries of the concept is far more problematic. Nor do mountains have determinate physical boundaries. 'Although the boundaries between the mountain and the air above its upper slopes may be determinate, prominent, and crisp, it is usually the case that, as we proceed downwards towards the foot of the mountain, no single candidate boundary is distinguishable at all.'[6] Introducing Jules Blache's *L'homme et la montagne* (1933), the French geographer Raoul Blanchard admitted the impossibility of a satisfactory definition of mountain.[7] And when forced to come up with such a definition (for example, in encyclopaedia entries), geographers have always done so reluctantly and not without a certain embarrassment.[8]

What is the difference between a mountain and a hill? Height, one might reasonably answer: mountains are taller than hills. But how much taller? In Christopher Monger's romantic comedy *The Englishman Who Went Up a Hill but Came Down a Mountain* (1995), two First World War English surveyors infuriate an entire Welsh village when they inform them that Ffynnon

Garw, their neighbouring 'mountain', is only a hill because it is slightly short of the required 1,000 ft (305 m) in height. This would result in its exclusion from the map, and, in the villagers' minds, in a possible modification of the Welsh boundary. The outraged villagers then resolve to add the missing height with their own hands and set out to pile soil excavated from their garden plots on the top of the hill. Eventually the hill is raised above the required height to qualify for the mountain label and thus be included on official Ordnance Survey (OS) maps.

The story is fictional, but it highlights the selectivity (and thus exclusivity) inherent in any mapping process and its perverse logic: only what is marked on the map counts, even though it is an arbitrary choice. This, of course, can have dramatic impacts on territory and on the lives of those who inhabit it (for example, setting or altering a boundary line). Ultimately the movie also reminds us of the artificiality and limits of any attempt to categorize landforms, including mountains.

Defining peaks through numbers, however, is no mere fiction. The U.S. Board on Geographic Names used to characterize a mountain as being 1,000 ft or taller, but this classification was abandoned in the early 1970s and nowadays the United States Geological Survey (USGS) concludes that the term does not have a technical definition in the U.S. By contrast, the standard British

Villagers add extra feet to their 'mountain' in Christopher Monger's *The Englishman Who Went Up a Hill but Came Down a Mountain* (1995).

OS classification still establishes a normative 2,000 ft (610 m) as the clear line between hill and mountain. Hence in 2008 Snowdonia's Mynydd Graig Goch, another Welsh hill, officially became a mountain after amateurs found it was 76 cm (30 in.) taller than previously thought – a contemporary re-enactment of Monger's movie.

Height is not the only criterion for defining a mountain. Geologists, for example, define mountains by their structure, rather than by their elevation above sea level. Hence, observes Lowell Thomas,

> some rugged highlands in plains and plateaus, such as those in Tibet, are certainly 'mountainous', but they are not really mountains. On the other hand, there are flat, low-lying rock surfaces in Canada and elsewhere – New York City, for instance – which are true mountains in the geological sense of the word. They are low now because they have been eroded to near base level, but they are still called mountains because of their underlying geological structures.[9]

Dictionary definitions, by contrast, tend to focus on visual characteristics. The *Merriam-Webster* dictionary, for example, emphasizes visibility, shape and the relationship with the surrounding environment: a mountain is 'a landform that rises well above its surroundings, generally exhibiting steep slopes, a relatively confined summit area, and considerable local relief'. Other definitions stress natural magnitude and the positioning in the landscape. According to the Macmillan dictionary, a mountain is 'a natural structure like a very big hill that is much higher than the usual level of land around it'. Similar definitions are found in other languages, including French and Arabic.

Such designations subordinate the naming to a specific landscape context and thus to direct human experience. This principle, however, is relative and in most cases culture-specific. 'What the inhabitants of cities such as Lyon or Turin would call "mountain" – the Alps, for example, as seen in the distance', observes historical geographer Bernard Debarbieux, 'might be a totally inappropriate designation for people living in the Alpine valleys', for whom 'mountains' are the high pastures and passes (rather than the summits).[10] Similarly the monks of Mount Athos, a monastic peninsula in northern Greece that culminates in a 2,033-m-high peak, call the upper part of the ridge 'mountain', but refer to the peak simply as 'Athos'.

Another way of defining mountains is by adopting a cartographic approach (as opposed to a landscape approach). This is a view from above, rather than from ground level. In this case, mountains appear less as isolated objects or breaks on the horizon than as parts of an integrated system, as tangible 'signs of an organizational principle of the cosmos'.[11] Aztecs called

View of the western slope of Mount Athos, northern Greece.

Al-Idrisi, *Mappa mundi*, 12th century. This map was originally drawn with South at the top; it has been inverted for the sake of clarity.

their territorial units *altepetl*, a composite term embedding the words for mountain and water, and imagined the whole world as a collection of such units.[12] On Islamic medieval maps, such as al-Idrisi's twelfth-century *Mappa mundi*, mountains feature as the sources of great rivers and as the backbones of Creation, as stated in the Quran: 'He hath created the heavens without supports that ye can see, and hath cast into the Earth firm mountains, so that it quakes not with you' (Sura 31:10). At other times, their shapes mimic wonderful Arabic calligraphy, so as to suggest the beauty and harmony of the Creation of which they are part. On Western Ptolemaic chorographic tables and world maps, a genre that developed in the fifteenth century, mountains likewise feature as solid backbones of continents and regions, as they do on traditional maps of Korea, a nation that has been historically defined by mountains.[13] In all these cases, mountains are subordinate to their function within a bigger organic whole pulsating with life: the earth, a continent, a region, a nation state.

These two ways of conceptualizing mountains – through direct experience and through global vision – do not require standardization. So when did mountains start to be measured and why? The mountain as a category of discrete, measurable objects is a product of a typically modern Western way of seeing and imagining the world less as an organism than as an container of landforms. In his multivolume *Encyclopédie, ou dictionnaire raisonné des sciences, des arts et métiers par une société de gens de letters* (1751–65), Denis Diderot defines mountains as 'large masses or inequalities of the earth that make its surface rough'.[14] The concept of altitude emerged in the seventeenth century, but was fully established only in the eighteenth. Determining the height of mountains became part of the larger Enlightenment project – the construction of a universal knowledge based on reason, measurement and classification. The

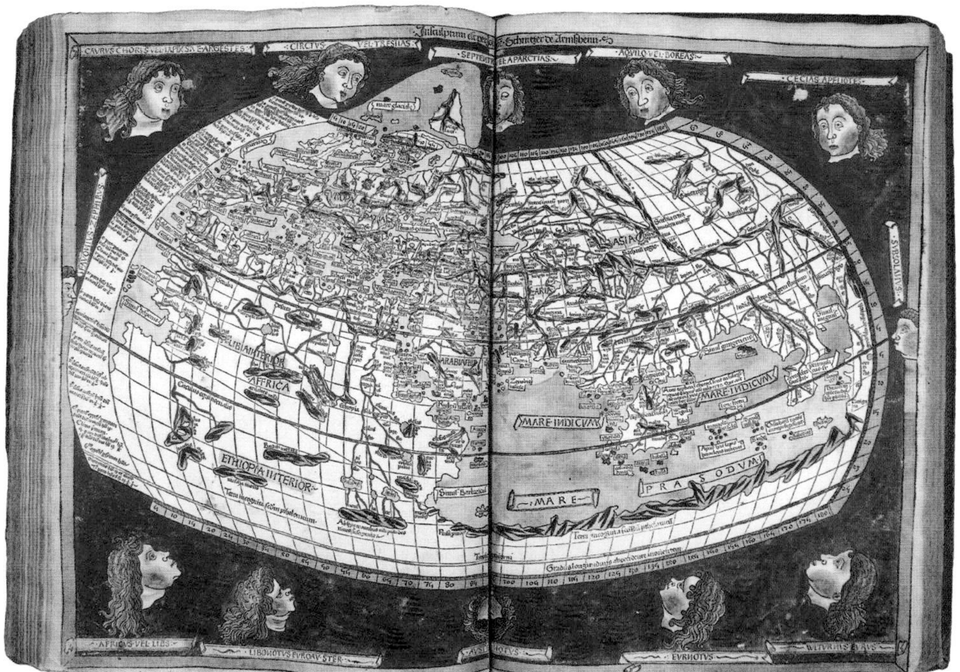

Johannes Schnitzer's Ptolemaic world map from the 1482 Ulm edition of Ptolemy's *Geographia*.

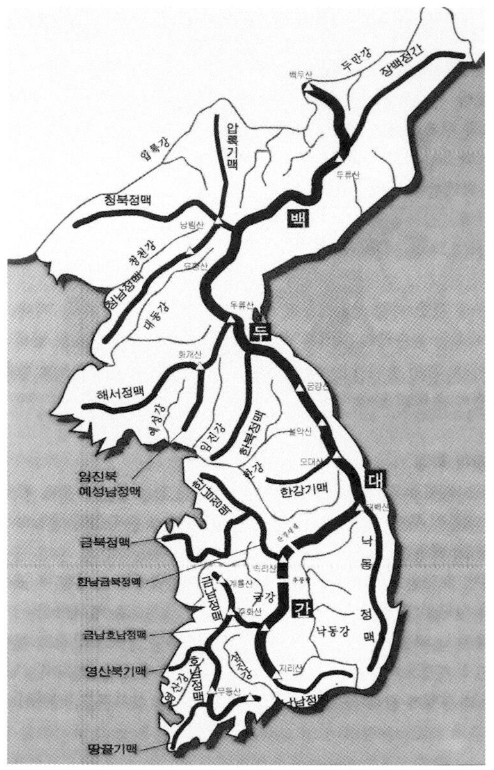

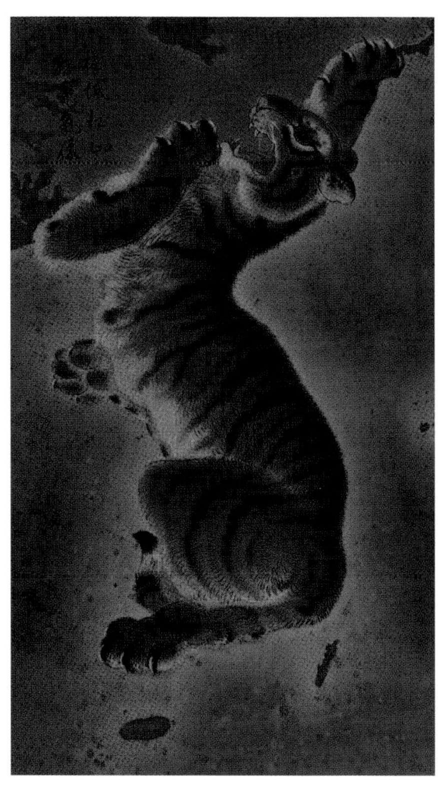

measurement of terrestrial summits developed alongside that of oceanic depths. Measuring mountains, Debarbieux observes, helped circumscribe and turn them into well-defined objects. A mountain was thus no longer thought of in terms of its context but of its own characteristics.

Representation of the restored traditional mountain system in Korea, based on an interpretation of ancient literature.

Early 20th-century zoomorphic map of Korea.

Altitudinal measure also allowed direct comparison between different types of mountains from all over the world based on an absolute, universal scale. In the nineteenth century memorizing mountain names and heights (along with the length of rivers) became part of geography students' curricula, as it continues to be in many primary schools around the world today. Comparative tables and global maps of the 'highest peaks' proliferated in textbooks and atlases that exalted 'the beauty brought to modern science by comparing facts of the same kind, by arranging them in classes and reducing them to general principles'.[15] Chimborazo, Everest, Kilimanjaro, Etna, Mont Blanc, even the Rock of Gibraltar, were thus taken out of their geographical context and grouped together in a grid or against a scale, often generating compelling imaginary landscapes. Direct visual comparison between familiar mountains (such as the Alps or Pyrenees, for example) and the newly explored giant peaks of Asia and South America helped memorization and at the same time reinforced Eurocentric notions of 'normality', as opposed to exotic 'otherness'.

With scientific measurement, the definition of the word 'mountain' itself changed. In the middle of the century, the German geographer Carl Ritter raised the problem of the vague meaning of the term. In 1873 the Austrian military surveyor and geographer Carl von Sonklar devised the first quantitative definition of the word 'mountain'. He based the distinction between hill and mountain on relief and set the threshold at 200 m from the valley floor to the summit. Some twenty years later, Albrecht Penck, another German geographer, defined a mountain as 'an area declining in all directions from a given location', thus confirming its status as a discrete object.[16]

W. R. Gardner, detail from the wall map in C. Smith's *Comparative View of the Heights of the Principal Mountains &c. in the World* (1820), engraving.

Over the following century new definitions came to encompass other aspects, including the botanical and the sociological. Scientific definitions reflected the specific topographic and topoclimatic conditions of mountain areas. For example, in 1942 Carl Troll, a geographer and botanist from Munich, based his definition of 'mountain' on landscape ecology criteria such as the altitudinal limits of different types of vegetation. Recent legal definitions, by contrast, tend to emphasize the human aspect. The Council of Europe, for instance, interprets mountain regions to mean 'areas whose altitude, sloping terrain and climate create special conditions which affect the pursuit of human activities'.[17]

Perito Moreno, Argentina.

Vernacular, cosmological and scientific definitions of mountain all reflect different ways of seeing and of coming to terms with creation: as experienced from within, or gazed on from a distance; as a living organism, or a collection of discrete objects; as an Other to be feared, or domesticated. This book's themes cover the holy and the diabolic, life and death, vision and time, science, technology and heritage. Two fundamental tensions underpin each of these themes: on the one hand, the tension between the circumscribable and the sublime; on the other, the tension between the awe and fear the latter generates. It is precisely through these tensions that mountains take shape and, in turn, have helped shape our environmental consciousness and our place in the world.

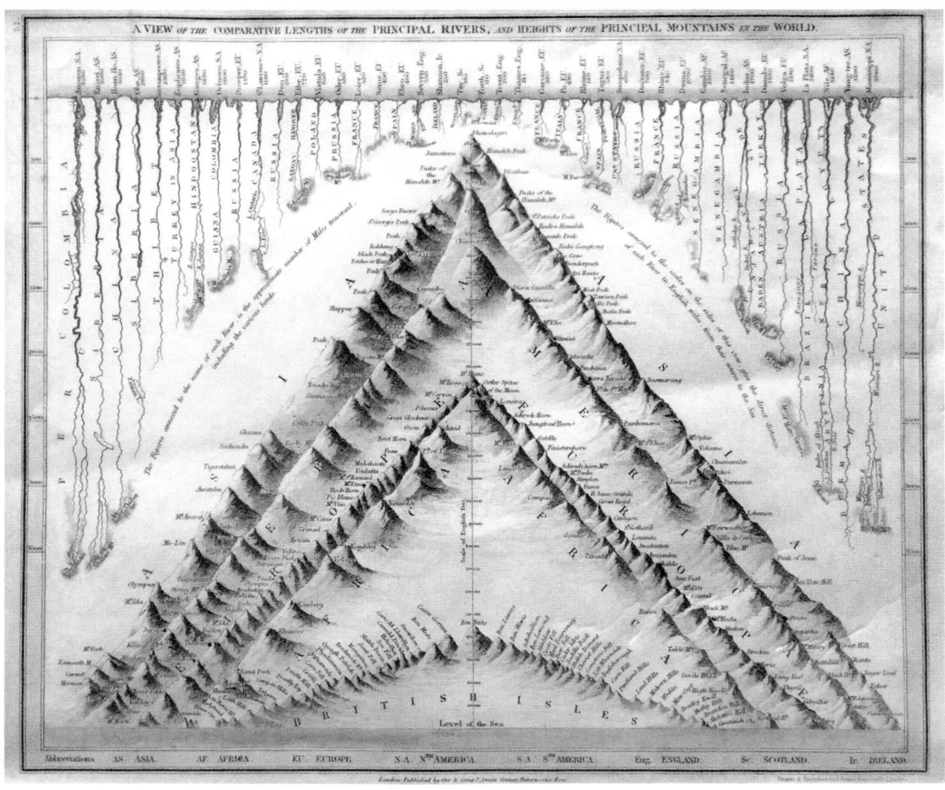

John Dower, 'A View of the Comparative Lengths of the Principal Rivers, and Heights of the Principal Mountains in the World', from *A New General Atlas of the World* (1844), engraving.

2 Mountains, the Holy and the Diabolic

The word 'holy' derives from the Germanic *halig*, which means something that must be preserved 'whole' or intact, something not to be transgressed or violated. The Greek word for 'holy', by contrast, is less about conscious preservation and more about unmediated response: *aghios* comes from the verb *azomai*, which means simply 'to stand in awe, or in fear'. The history of holy mountains rests on the tension between these two meanings – respectful stewardship and speechless terror, separation from the ordinary and interaction with the transcendent. It is through this act of separation, or rather, through the sudden encounter with the unexpected, with the utterly different, that humans are called to turn to their most inner self, and to the divine.

'Places that are strange and solemn strike an Awe into us and incline us to a kind of superstitious Timidity and Veneration', wrote the seventeenth-century British theologian Thomas Burnet.[1] Take the cones of Kilimanjaro, Mount Fuji, Mount Athos or Croagh Patrick: they all cause unexpected breaks on the flat horizon. They take us by surprise. Or take the imposing summit of Machu Picchu, the otherworldly rocks of the Sinai massif at sunset, or the sandstone pinnacles of Meteora as they cast their dark shadows on the plain of Thessaly: confronted by such sights, we find ourselves lost for words.

In the Abrahamic traditions God usually chose to speak through charismatic prophets and holy men and through equally charismatic places set apart from the inhabited world – in other words, through wilderness, and especially through mountain

Jean-Marie Saint-Eve, *Temptation on the Mount*, 1854, watercolour.

wilderness. Elijah encountered God in utter silence on Mount Horeb; Moses, on the summit of Sinai, in a thick darkness, beyond vision and human comprehension. Christ disclosed his divine nature to his three disciples in shining vests on the top of Tabor. Muhammad received his first revelation from the Archangel Gabriel in a cave on Jabal el-Nur, 'the Mountain of Light', near Mecca, and ascended to heaven from Mount Zion in Jerusalem, where the Dome of the Rock stands today. Like the prophets of the Old Testament and Jesus himself, Christian solitaries have continued through the centuries to seek refuge from society in the howling wilderness of the high places around the Mediterranean and beyond.

The sacred, Edwin Bernbaum observes, does not merely present itself to our gaze; 'it reaches out to size us in its searing grasp'.[2] And so do mountains. Their sheer size, their charisma or simply their 'difference' from the surrounding environment have long captured the attention of different cultures and anchored the beliefs of most religious traditions, both monotheistic and pagan. Different types of mountains around the world enshrine different credos, values and aspirations. 'The graceful cone of Mount Fuji has come to represent the quest for beauty and simplicity that lies at the heart of Japanese culture. Mount Everest stands out, even in the modern, secular world, as an inspiring symbol of the ultimate.'[3] Less impressive prominences have enjoyed no lesser importance. For the Romans, the Palatine Hill, standing a mere 40 m over the Forum Romanum, was the centre of the world, just as the hill of Zion was for the Hebrews.

Because of their inhospitality, in certain traditions mountains have also provided privileged abodes for demons. Historically, for example, Alpine and Scandinavian summits were by no means holy places, nor, until the nineteenth century, were they the romantic settings we appreciate today. They were inhospitable, unproductive lands and the sinister haunts of dragons, trolls and evil spirits. It was not until they were climbed and measured that those lofty peaks were exorcized.

Different mountains articulate different sacred geographies; they can act as networked landmarks, as centres and boundaries, as ladders to heaven, as thresholds to the unknown and places of fear. No matter what their exact height or geographical location, mountains help us pin down the ungraspable and utter the unspeakable. Solidly anchored on earth yet stretching to heaven, they bridge what we see and what we do not see. They are material as much as symbolic realities, charismatic features of the land and of the soul – for geography is, after all, simply 'a visible form of theology'.[4] This chapter offers a brief journey through sacred peaks and explores how different mountain geographies have contributed to structuring different cosmological views in various civilizations around the world and throughout history.

Networks and landmarks

In many cases, high places have literally shaped visions of the world and of the universe. Ancient Chinese peaks, for example, inspired the Taoist principle of *yin-yang*. Originally the two words *yin* and *yang* referred to the shaded and sunlit sides of a mountain. Over time, they came to designate 'the complementary opposites whose union creates the world – darkness and light, moisture and dryness, female and male, nonexistence and existence'.[5] These cosmic forces still meet in the high places of the Middle Kingdom. In traditional Chinese culture mountains are deemed to be the 'bones' of the earth and of nature, the sources from which all places originate and where *yin* and *yang* alternate. 'A single mountain combines several thousand appearances', an eleventh-century Chinese landscape painter claimed, referring to their combinations of vital energies and ever-changing configurations.[6] A ninth-century poem describes mountains' various shapes in terms of the hexagrams of the *I Ching* or *Yi Jing* (*Book of Changes*):

> Some, like the omens cracked in tortoise-shell,
> Some, like the hexagrams, divided into lines.
> Some are like Bo, stretching across up the front,
> Some are like Gou, broken in the back.[7]

With their curiously shaped limestone pinnacles and granite peaks, grassy slopes and fault-block eminences beautified by an occasional temple or gnarled pine, Chinese high places are dynamic sites in which the eternal renewal of the universe is revealed to humans.

The regular mountain- and hill-range patterns in Korea likewise allowed the ancient geomantic principles of Chinese feng shui (*pungsu* in Korean) to be adopted from architecture and applied to the identification of ideal landscapes for burial. Traditionally Koreans did not bury their dead in cemeteries, but in places thought to be auspicious. Shapes of great constellations known as the Celestial Animals were read in land patterns that were believed to channel special energies and influence the luck and prosperity of the descendants of the deceased. Hence specific conformations of mountain and hill ridges deemed to reflect these cosmic patterns continue to be dotted with graves.

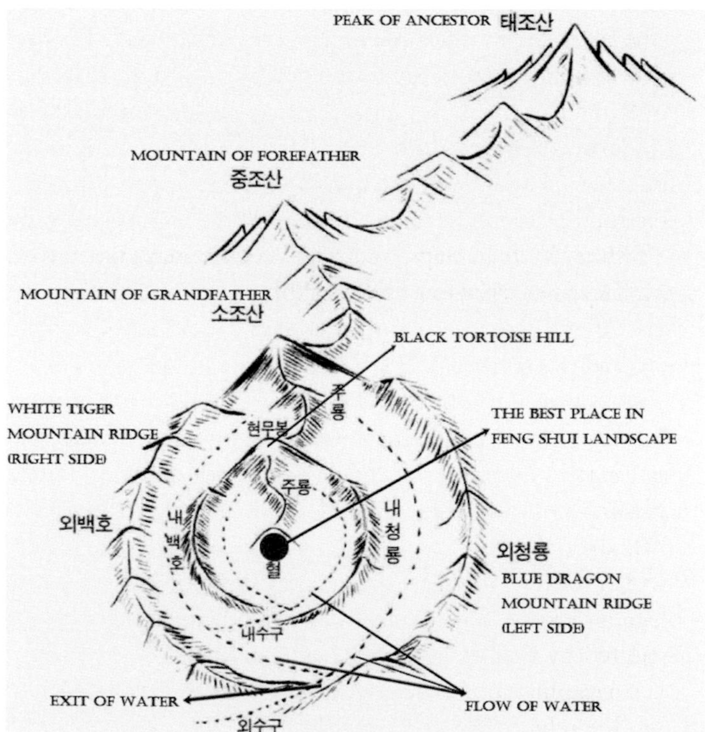

Drawing of the ideal Korean *pungsu* pattern, 2003.

Disputes over auspicious grave sites were a common occurrence during the five centuries of the Joseon dynasty (1392–1897). Today in both North and South Korea vast amounts of money are invested by politicians, wealthy citizens and expats to secure appropriate burial spots for themselves and their families. The phenomenon has reached such proportions that environmentalists have expressed serious concerns, as hillsides are razed to create propitious patterns and others are increasingly crammed with elaborate grave sites.[8]

If Korean sacred microcosms are defined by harmonious ridges and hilly patterns not very different from one another (and thus replicable throughout the country), in other cultures they are marked by holy peaks that stand out of the landscape. Such heights act as powerful landmarks for both physical and spiritual orientation. Located hundreds of kilometres apart, Mount Taylor in New Mexico, the San Francisco Peaks in Arizona and Blanca Peak and Hesperus Mountain in Colorado set the borderlines of the Navajo ancestral tribal land. They also set the structure of the Navajo cosmic scheme. To each of these four mountains the Navajo associate one of the four cardinal directions, a time of the day and a colour. Blanca Peak, for instance, is associated with dawn and white and Hesperus with night and black.

Unlike Western maps, traditional Navajo maps are made of sand. They are carved out of the earthly matter which they represent. Here Blanca and Hesperus are marked respectively by a white shell and a jet stone, whereas Taylor and the San Francisco Peaks are represented by a turquoise and an abalone shell.[9] These ephemeral maps are used in rituals to restore cosmic order each time a natural disaster occurs. Chants and dances follow the clockwise motion of the sun as it advances across the sky and moves over the four peaks. The image of a circular cosmos bounded by the four sacred mountains is central in Navajo symbolism and today it continues to appear on their seal and flag. Separated by vast distances and transiently brought together in cartographic space, the four Navajo peaks articulate sacred space. The schematic interconnection between these mountains

Mists and clouds on Huangshan in Anhui province, eastern China.

continues to shape the boundaries of an imagined community, even if the peaks are not in sight of one another.

The Navajo Great Seal featuring the four sacred peaks.

In the ancient Aegean, by contrast, sacred mountain networks were based precisely on their visual interconnectedness, though at a much more intimate scale. The Minoan religion, which flourished on Crete between 3000 and 1500 BC, centred on the cult of the 'mountain mother', a female goddess associated with the fertility of the land and worshipped on high places. Neither too high to be forbidding nor too low to pass unobserved, Cretan peaks offered relatively easy points of access to the divine. The more than fifty mountaintop shrines present on the island could be reached on foot within three hours from the main settlement and they were set in sight of each other, as well as in sight of the villages underneath. Large sacrificial bonfires were lit as part of ceremonies, providing spiritual comfort to the villagers in the valley. During festival nights a network of sacred beacons linked various regions and offered the faithful the possibility to undertake a 'visual pilgrimage' through these holy peaks.[10]

In classical Greece mountains became part of broader sacred networks that extended to both the mainland and the islands. The Greek landscape, the archaeologist Vincent Scully notes, 'is defined by clearly formed mountains of moderate size, which bound definite areas of valley and plain. Though sometimes cut by deep gorges and concealing savage places in their depth, the mountains themselves are not horrendous in their actual size.'[11] Majestic peaks like Olympus, towering over the plain of Thessaly, or the pyramidal cone of Athos arising from the flat surface of the Aegean, are thus exceptions to this harmonious relationship between valleys and mountains. They signal ruptures on the visual horizon and in the ordinary. In Greece, as elsewhere, such anomalous contrast with the surrounding land made these peaks sites of awe, and even the dwellings of specific divinities. As the highest summit, snow-capped Olympus (2,917 m) was the seat of Zeus and the twelve gods. Further south, the gentler slopes of Parnassus (2,457 m) were deemed to host the Muses. Sanctuaries of Zeus and Apollo were built in dramatic high places, such as Mount Athos. In many of Apollo's sites the abstract, mathematical order of the Greek temple was made 'to contrast most sharply with the rough masses of the earth, dramatizing at once the terrible scale of nature and the opposing patterns which are the result of disciplined human action in the world'.[12]

Ancient Greek sacred geographies were closely intertwined with the open-air geographies of coastal navigation. The word for mountain, *oros*, also means landmark, for such was the function of Greek high places.[13] Mountains and promontories acted as essential landmarks for sailors and were often topped by temples. These signalled sites of severe storms, as the high reliefs on which they were located often converted the straits and bays below into sea canyons through which winds blew seasonally with restless violence.[14] Sometimes holy peaks served as beacons. Other times, they functioned as meteorological centres: 'a mantle of clouds about Athos presaged a storm; a girdle of clouds half way up its slope indicated a southerly wind and eventual rain.'[15] Temple-topped mountains and promontories marked the fine line between known land and boundless sea. They were the first familiar features the seaman would have glimpsed on return to his homeland.

Axes mundi

While Koreans, Navajo, Minoans and ancient Greeks built their sacred spaces and world images through mountain networks, other cultures have been dominated by the overwhelming shadow of a single peak, or what the philosopher and historian of religion Mircea Eliade called the *axis mundi*, that is, 'the centre of the world'.[16] The New Zealand Taranaki tribe, for example, centre their existence around the mountain of the same name, which rises dramatically on the North Island for 2,518 m. As the origin of the rivers that pour into the valley, the snow-capped mountain is also envisaged by the Taranaki as the source of life. A local legend roots the origins of the tribe in the love between Mount Taranaki and a neighbouring volcano and a titanic battle with a rival peak. Life thus comes from their mountain, and when life is taken away, the Taranaki are ultimately returned to the mountain.[17]

Similarly, the Chagga tribe who live on the slopes of Kilimanjaro regard the mountain as the centre and source of their lives and draw sustenance from the snows on its top, the highest point of Africa. Even after their conversion to Christianity, the mountain has continued to dominate the Chagga's spiritual lives and everyday practices. For example, most of their churches have their altars on the side closest to Kibo, the highest of the three volcanic cones of Kilimanjaro. Homes and shops are likewise built next to the Chagga's farms on the side nearest to Kibo. Sleeping with the head towards the mountain and spitting in the direction of Kibo before leaving on a journey are common propitiatory practices among the tribe.[18]

If the dominance of Mount Taranaki and Kilimanjaro in the lives of the Taranaki and Chagga is intimately linked with their daily experience, other cultures set invisible mountains or mountains out of physical reach as the pivots of their universes. Hindu and Buddhist images of the cosmos, for example, are centred on and supported by Mount Meru, a mythological summit believed to be 1,082,000 km high – 85 times the earth's diameter. On medieval mandalas, such as the fourteenth-century Yuan dynasty silk map preserved at the Metropolitan Museum of Art in New York, Meru features in the centre as an inverted pyramid topped by a lotus, a Buddhist symbol of purity. Close to its base are the sun and the moon. Around the mountain, in different colours, are the four continents of Indian mythology, each of which contains peaks marking the cardinal directions.[19]

Gotenjiku Zu, or 'Map of the Five Indies' (1364), the oldest Japanese world map, is based on an earlier Chinese map and presents an egg-shaped earth likewise centred on Meru but

Cosmological mandala with Mount Meru, Yuan dynasty (1271–1368), China, silk tapestry.

entirely dominated by mountains. India, Buddha's native land, occupies the central part of the map, whereas countries including China and Japan are cast to the periphery. Symbolic features are identified with actual places; Meru, for example, is associated with Mount Kailash in Tibet.[20] If mountains are the places 'from which all places originate', in this image Mount Meru appears as the origin of all the mountains. This iconography echoes the South Asian belief that a great number of holy mountains in that region are indeed fragments of Mount Meru that have been bestowed on the local inhabitants by the gods to bring them closer to the centre of the cosmos as the source of stability and blessing.[21]

According to Buddhist belief, however, the universe we know (which is represented on these maps) is just one amid millions of others through which souls transmigrate during successive incarnations. Each world is layered vertically atop Meru and ascent to a successive plane is based on moral conduct. This vertical hierarchy culminates in the non-form, or *nirvana*, the ultimate stage to which each soul should tend. Identical maps featuring the layers supported by Meru are engraved on the lotus petals on which sits the giant eighth-century bronze statue of the Buddha Vairocana at Todai-ji in Nara, Japan. At the bottom of each petal circular representations of the earth are enclosed within another seven petals. In the centre stands Mount Meru, surrounded by rings of mountain ranges and water, and topped by 25 horizontal strata of the heavens containing small Buddhas and palaces.

This cosmic geography operates at a variety of scales, including that of the human body. South Asian yoga practitioners often visualize the universe as one with their own organisms and identify the cosmic axis of Mount Meru with 'a channel that rises from the base of the spine to the top of the head'.[22] Human

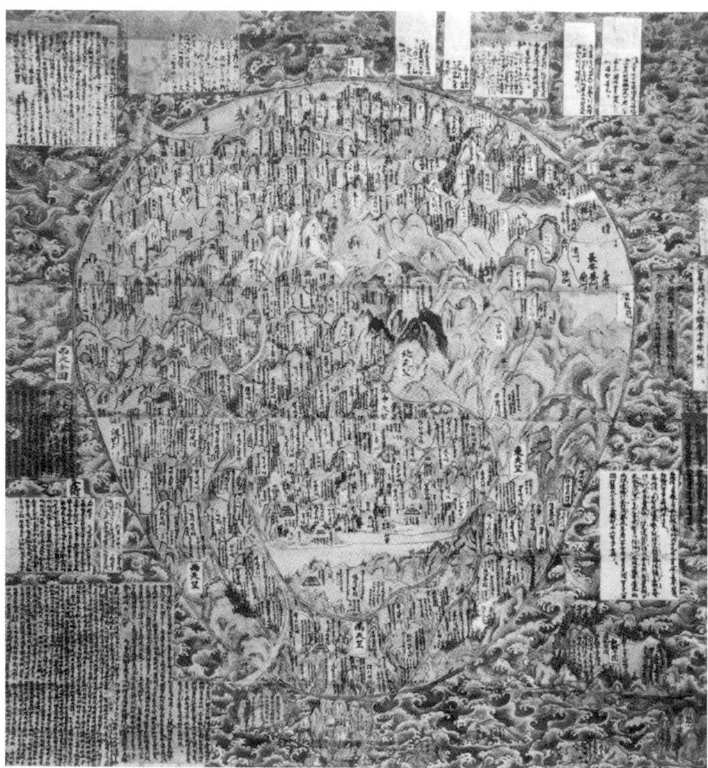

Gotenjiku Zu, or 'Map of the Five Indies', Japanese world map, 1364.

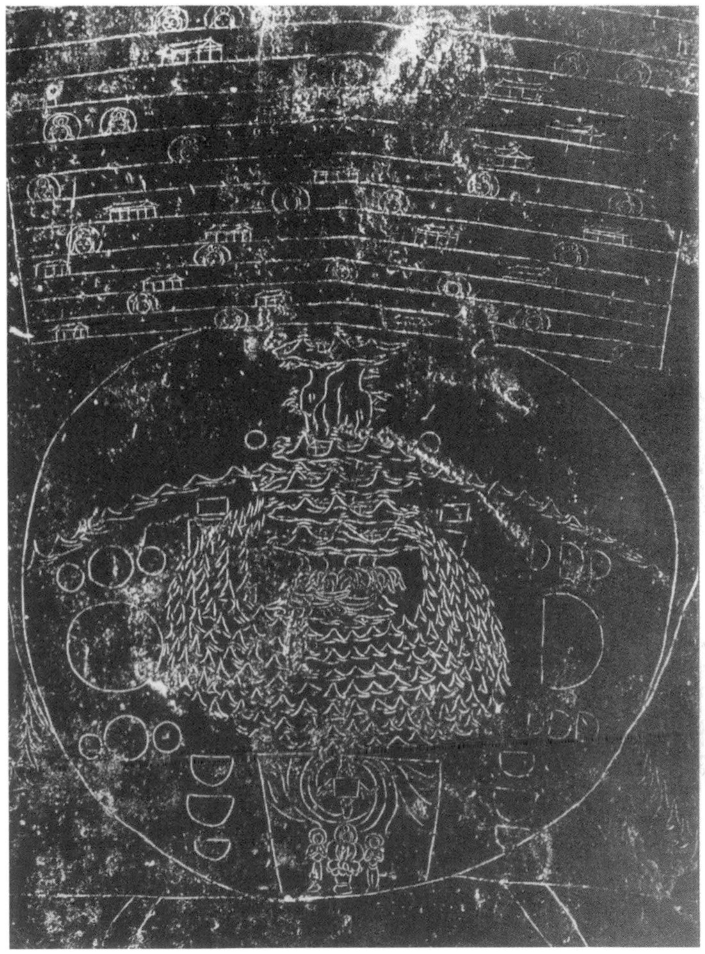

Oblique view of Mount Meru and the universe engraved on the bronze pedestal of the statue of the Buddha Vairocana in the Todai-ji temple in Nara, AD 749.

bodies thus become maps of the same Meru-centred cosmos of which they are part.

The belief that mountains are the places nearest to heaven and thus cosmic *axes mundi* is shared by other religious traditions, including the Judaeo-Christian. In the earliest Hebrew cosmology God was believed to reside in the Upper Chambers above the firmament (Psalm 104:13). He was therefore always imagined as coming down to meet with the faithful, giving mountains a special significance as being closer to His dwelling place. As such, mountains are one of the most present

geographical features in the Bible. Unlike in the Chinese tradition, however, in Judaism and Christianity not all of them are holy, nor are they intrinsically holy. Biblical holy peaks were sanctified through theophanies, that is, through place- and time-specific manifestations of the divine (for example, Moses' encounter with God on Sinai, or Christ's transfiguration on Mount Tabor). Consecrated to Old Testament prophets and physical outposts of the stations in the life of Christ, biblical holy mountains are not necessarily striking (some of them are mere hills), yet they act as powerful anchors of tradition, as well as ladders 'uniting heaven and earth . . . the point at which the transcendent might enter the immanent' – in other words, as *axes mundi*.[23]

Cosmas Indiocopleustes' tabernacle-shaped cosmos, from an 11th-century manuscript of the *Topographia christiana*, in the Holy Monastery of Saint Catherine.

The most stunning Judaeo-Christian mountain is not found in the Bible, but in a three-dimensional diagram contained in

Cosmas Indicopleustes' *Topographia christiana*, a sixth-century illustrated cosmographical treatise. The author, a traveller-geographer from Alexandria who later in his life became a monk, imagined the cosmos in the shape of a tabernacle dominated by a huge peak. On Sinai, Cosmas believed, God disclosed to Moses 'an interlocking correspondence of spiritual and terrestrial geographies', which the monk literally mapped out in his treatise.[24] The higher part of the structure was occupied by the vaulted heavenly chamber inhabited by God. The rectangular lower prism enclosed the terrestrial realm.[25] This contained a rectangular *ecumene*, or inhabited world, surrounded by the ocean and overlooked by a mountain 'as high as the breadth of the land towards the northern and western regions' and beyond which the sun, carried by two angels, disappeared every evening.[26]

The huge cosmic mountain dominates the *ecumene* to the same extent that the presence of Mount Sinai, the mountain of Moses and the holiest Old Testament peak in Eastern

Saint Catherine's Monastery at the foot of Mount Sinai.

Mount Tabor, the mountain of the Transfiguration, Israel.

Christianity, dominated Cosmas' imagination. Craggy, barren, awe-inspiring, even though hardly indistinguishable from the several other surrounding peaks, Sinai is the mountain of the Covenant and the Law. On the summit of this 2,288-m massif, located in one of the most arid and inhospitable regions of the world, Moses received the Commandments 'in a thick cloud' (Exodus 19:9; 1 Kings 19:8–13). Byzantines such as Cosmas saw in Sinai a foreshadowing of Tabor, the New Testament theophanic mountain par excellence. Rising to a mere 575 m, unlike Sinai, the lofty hill pops up almost unexpectedly from the flat landscape of Lower Galilee and the visual contrast is enough to justify its biblical epithet of 'high mountain' (Matthew 17:1). Besides their different altitudes and visibilities, the two mountains also present very different environments. In the sixth-century Sinai was portrayed by the historian Procopius as a wild and rugged mountain. By contrast, the Armenian pilgrim Elisaeus in the seventh century portrayed Tabor as a delightful *locus amoenus* surrounded by springs and by trees producing 'all kinds of sweet fruits and delightful scents'.[27]

The juxtaposition of Sinai and Tabor has been exploited through the centuries to illustrate the dual nature of Eastern

Theophanes the Greek, *Transfiguration*, 14th century.

Christian doctrine. Sinai symbolizes its 'aniconic power', whereby God is defined as uncreated, boundless, unattainable, thus inviting us to recognize our 'knowledge about unknowledge'.[28] On Mount Sinai Yahweh revealed Himself to Moses in a scarcity of images. Like all mystic seers, Moses had to enter into the dark cloud of unknowing. Tabor, the Mount of Transfiguration, by contrast, symbolizes the iconic, imaginative power of Christianity, the approach through which we comprehend the natural world and translate it into knowledge of the divine.[29] On the top of the gentle, verdant Mount Tabor, Jesus of Nazareth was transfigured before Peter, James and John. 'His face shone like the sun, and his clothes became as white as the light' (Matthew 17:2). What was not given to see and know to Moses was made manifest in the New Testament.[30]

In medieval Byzantine iconography Tabor, the Mount of Transfiguration, is usually transfigured into the mountains of the Old Testament's revelations: Mount Horeb (topped by Elijah, who stands for 'the Prophets') on the left and Mount Sinai (topped by Moses, who symbolizes 'the Law') on the right

Post-Byzantine fresco of the Transfiguration painted on the upper part of the southern *choros* in the main *naos* of Docheiariou's katholikon, Mount Athos, 1568.

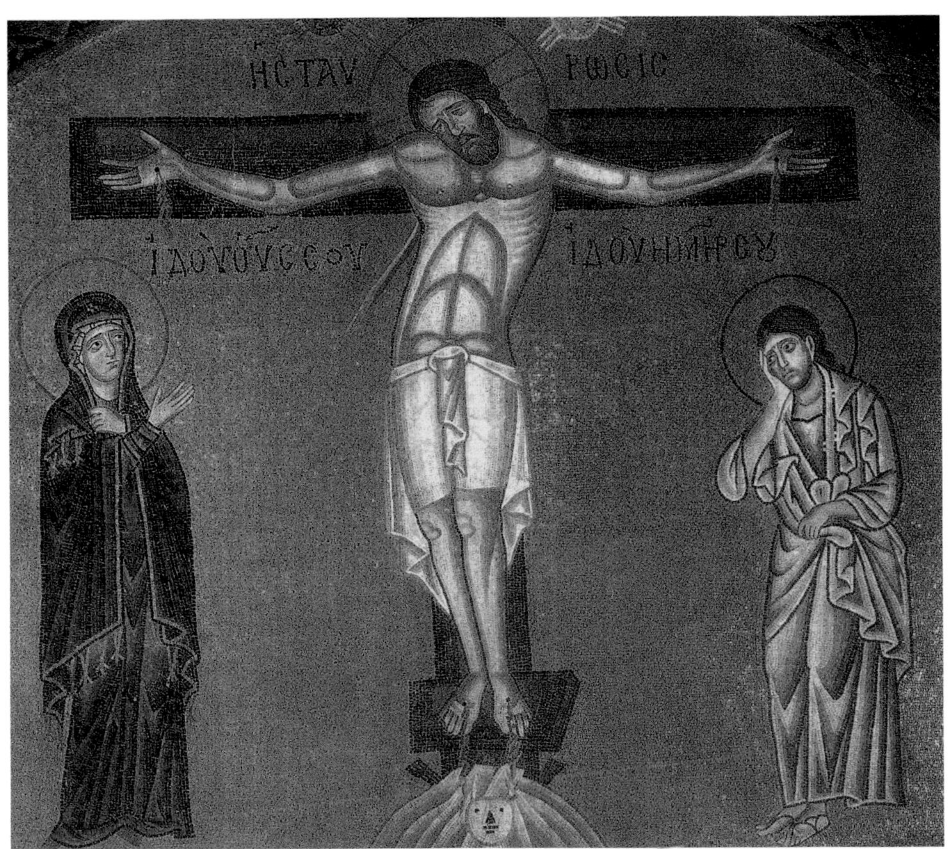

The Crucifixion, featuring the skull of Adam inside Golgotha, Hosios Loukas monastery, mosaic, 11th century.

(Matthew 5:17). In the upper part of the icon opposite, the two prophets mirror each other, as they engage in a timeless dialogue with a motionless Christ dressed in glistening, intensely white clothes and enclosed in a mandorla of light sometimes reminiscent of the concentric Aristotelian cosmos – the fulfilment of both the Law and the Prophets. Tabor is superimposed over the other two mountains. Old and New Testament narratives converge in space and in time in what some have defined a 'Christian mandala'.[31] On Tabor, the 'navel' of the cosmos, past, present and eternity converge in a single moment. Likewise, the cosmographic and topographic scales converge in a single scene – and in the figure of Christ, the centre of the universe.

Hereford *mappa mundi*, *c.* 1285.

In the New Testament Mount Tabor foreshadows another mountain: Golgotha, or the 'place of the skull'. This small hill, located just outside of the walls of Jerusalem, was believed to be the burial place of Adam. Unlike Tabor, Byzantine iconography usually downplays the size of Golgotha. On Byzantine and post-Byzantine icons of the Crucifixion, a small cave at the bottom of Golgotha shelters the skull of the progenitor of

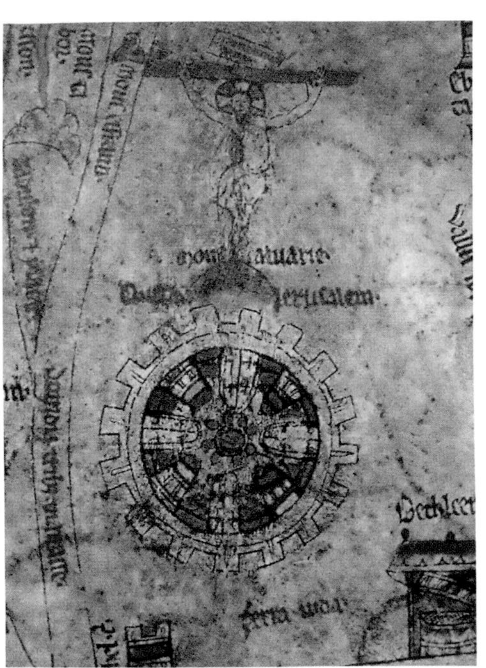

Detail of the Hereford *mappa mundi* featuring Christ's crucifixion at the centre.

mankind, now part of the dust of the earth from which he was formed. The size of the hill nearly disappears under the size and weight of the cross, as if to highlight the function of the latter as a link between earth and heaven. Through his sacrifice, Christ, the New Adam, redeems the sins of the old one and creates a new *axis mundi* – an axis of salvation between death and eternal life. On this axis the Latin West centred its world image. On *mappae mundi*, or Western medieval Christian images of the world, the three continents converge on Jerusalem and universal history in the moment of Christ's crucifixion.

As Christianity was exported to the New World, new holy mountains and *axes mundi* emerged from old vernacular ones. Spanish missionaries thought that placing their Christian saints at sites formerly occupied by indigenous deities would help the conversion of local populations. This practice, however, paved the way to syncretism. One such example is the Cerro of Potosí in southern Bolivia. Overlooking the town of Potosí, at 4,090 m (one of the highest in the world), the Cerro was popularly believed to be 'made of' silver and its mines provided a major supply for Spain. Because of its richness, indigenous populations associated the mountain with Pachamama, the Andean goddess of *tierra* (earth, soil), and worshipped it as a queen. With the advent of Christianity, Pachamama was naturally conflated with the Virgin. In an eighteenth-century representation by an anonymous painter, the Virgin is literally embodied in the mountain and topped by the Holy Trinity.[32] The painting gives visual expression to the words of the seventeenth-century Augustinian friar Alonso Ramos Gavilán, who compared Mary to a 'divine mountain'. As with other Christian *axes mundi*, the Virgin-mountain

Anonymous, *La Virgen María y el Cerro Rico de Potosí*, 1740.

joins earth to heaven, but, more characteristically, she also joins Christian and pre-Columbian traditions and beliefs.

Thresholds and invisible mountains

Mountains signal the presence of something different: 'something that fascinates, terrifies, attracts and threatens, all at once'.[33] As such, mountains have traditionally served not only as *axes mundi*, but as liminal spaces marking the boundaries between the known and the unknown – between our world and other worlds. Symmetrical ridges mark the boundaries of Korean auspicious sites, preventing the dispersal of cosmic energies. The four sacred peaks of the Navajo mark the boundary between the tribe's ordered universe and the threatening world outside, just as the Pillars of Hercules (the Rock of Gibraltar and Jebel Musa) signalled to ancient Mediterranean people the limits of their familiar world and the beginnings of the terrifying Okeanos, the river of chaos surrounding the earth.

In ancient Egypt the horizon was marked by the mountains of sunrise and sunset, the gateways to the netherworld. The Mesopotamian horizon, too, was bound by mountain-thresholds and associated with the Sun-god, Utu-Šamaš, who scaled the eastern mountains in his daily ascent on a stairway of lapis lazuli.[34] Much like the Navajos' tribal land and the Greek *ecumene*, the mountains defined the edges of the Mesopotamian known world, beyond which extended a mythical terra incognita.

Holy peaks, however, can and do also act as vertical thresholds, as the upward limits of this world. Traditionally they have thus been the preserve of few adepts: prophets, holy men, high priests, sages, shamans. 'Draw not nigh nither', says the Lord to Moses. 'Put off thy shoes from off thy feet, for the place whereon thou standest is holy ground' (Exodus 3:5). Only Moses was allowed to ascend the peak of Sinai; the rest of his people had to camp at the foot of the mountain.

Medieval Chinese mountains were likewise the preserve of charismatic solitaries. These were not countercultural backwoods trolls, but 'well-educated scholars and officials who renounced participation in public office for political and philosophical reasons' and chose to devote their lives to meditation.[35] Mountains, the eighth-century Chinese poet Liu Yuxi wrote, 'are famous not because of their heights, but because of transcendent lives', meaning both wondrous nature and spiritually elevated humans, beyond the reach of ordinary people.[36] It was possible, however, for ordinary people to gain contact with such high places and their spiritual benefits through miniature replicas, such as columnar rocks or incense burners, which, through

Bronze incense burner featuring sacred mountains, Eastern Han Dynasty, AD 25–220.

their convoluted shapes, conveyed the dynamic, erupting spirit of the place.[37]

Over the past century, with sacred boundaries becoming more tenuous under the pressure of international tourism, native populations from different parts of the world set claims to regain their hallowed grounds. Today no longer awe and fear, but fences, signs and regulations keep tourists, excursionists and sportsmen off the sacred sites around Uluru in Australia or in Devil's Tower in Wyoming, a peak sacred to twenty indigenous tribes, as well as to thousands of mountain climbers from all over the world.[38] In Japan ascetic climbers to the top of the holy mountain Tateyama share their sacred place with a booming ski resort.[39]

Besides the innumerable holy 'mountain gateways' around the world, a plethora of invisible mountains have populated the maps and geographical imaginations of many. Putting a mountain on a blank space on the map, it seems, has a reassuring effect. Hence on medieval Islamic *mappae mundi* one of the features that immediately grabs the attention for its prominence is the legendary Jebel al-Qamar, or Mountains of the Moon, from which, until modern times, the Nile was believed to originate. Located at the centre of a largely blank African continent, the mountains are the largest feature on the map and mark its south-up orientation.

Likewise, medieval Islamic maps of northwestern Africa and Spain always include Jabal al-Qilal, a mysterious triangular mountain-island off the Strait of Gibraltar at the mouth of the Mediterranean – 'a symbol for the Pillars of Hercules of yore adopted into the Muslim cartographic repertoire as a warning to readers of the dangers that lay beyond it'.[40]

Islamic cosmology set another phantom mountain island on the North Pole. Mount Qaf was deemed to be the farthest point on earth and to be made of emerald, which was thought to give the sky its blue tint. As with Dante's ascent of Mount Purgatory or Buddhists' ascent of Mount Meru, the ascent of this marvellous mountain marked spiritual progress. Alternative interpretations envisaged Qaf as a whole mountain range

Istakhri's map of North Africa and Spain, with the mythical Jabal al-Qilal featuring as a triangular island in the Straits of Gibraltar, 1173.

encircling the earth and comprising 'many great peaks, each millions of miles distant from the other'.[41] This interpretation is tied to Sufi works, such as the 'Conference of the Birds', a twelfth-century Persian allegorical poem narrating the story of birds in search of their mythical leader who withdrew to the Mountains of Qaf to contemplate. As the historian of Islamic cartography Karen Pinto suggests, these mountains 'reinforce the edge of the world and take on a sense of mountains of wisdom and contentment'.[42] From the fourteenth century this idea assumed visual form and started to become a standard feature on Islamic world diagrams. Here the Qaf range features as a scalloped outer band encircling the ocean and the planet, a boundary beyond which 'souls seek peace and unity with the maker and paradise waits'.[43]

A 'scientific' version of Mount Qaf is found on Western maps as late as the seventeenth century. Cartographers used the Rupes

The world map of Ibn al-Wardi, 1593.

Nigra et Altissima, an imaginary highest black mountain island set precisely on the North Pole, to explain why all compasses pointed to this location. 'In the midst of the four countries is a Whirl-pool, into which there empty these four indrawing Seas which divide the North', Gerhard Mercator wrote to John Dee.

> And the water rushes round and descends into the earth just as if one were pouring it through a filter funnel. It is four degrees wide on every side of the Pole, that is to say eight degrees altogether. Except that right under the Pole there lies a bare Rock in the midst of the Sea. Its circumference is almost 33 French miles, and it is all of magnetic Stone.[44]

On his map of the North Pole, the Flemish cartographer sets the Rupes Nigra at the centre of a concentric pattern of waters and quadripartite lands uncannily reminiscent of Buddhist mandalas.

As liminal places, out of the physical reach of most, mountains have also been privileged candidates for the location of earthly paradises. In the ancient Chinese tradition a mythical mountain above the barren heights of the Kunlun range in Tibet, which climbs up to over 7,000 m, was said to host a magnificent palace of jade surrounded by ramparts of gold. An unsurpassable river encircled the base of the mountain, cutting off all access to mortals.[45]

Likewise, various Church Fathers imagined the Garden of Eden on an inaccessible peak, far beyond the reach of mankind. For St Jerome (*c.* AD 347–420) Eden was a mountain

> hard to climb and amazingly high and in natural form like a high tower with the steep part as if it had been cut by hand. The way round takes more than one day. On the sides of the mountain trees are scarce. Many birds of various kinds fly round the mountain in flocks, but the mountain itself would seem to be without plants or moisture, and is far from any living growth in the desert.[46]

Three centuries later John of Damascus (*c.* AD 676–749) simply located Eden 'in the Orient, in the most elevated region on earth'.[47] The Church Fathers, however, were more interested in paradise as a condition of the soul than as a physical place on earth. Ephraim the Syrian (*c.* AD 306–373) thus described Eden as a conical mountain simultaneously transcending the world and enveloping human spirituality, as a feature outside space and time. The holy man declared to have gazed upon the mountain 'with the eyes of his soul'. Mount Eden, he wrote, rises above and encompasses the entire physical world:

> The summit of every mountain
> is lower than its summit,
> the crest of the Flood
> reached only its foothills;
> these it kissed with reverence
> before turning back
> to rise above and subdue the peak
> of every hill and mountain.
> The foothills of Paradise it kisses,
> while every summit it buffets.[48]

For Ephraim, Mount Eden is higher than any other earthly feature, and at the same time it includes all of them. Its circular base surrounds the entire creation, 'resembling that halo which surrounds the moon . . . having both sea and dry land encompassed within it'.[49] Ephraim's cosmic mountain is a reminder of the sacramental character of the material world and at the same time of the inner spiritual world that resides in the heart of each human. This interior realm, Ephraim argues, can only be revealed through symbols – and mapped through symbolic topographies.

In his *Life of Moses* St Gregory of Nyssa (*c.* AD 335–395) charted the journey of the soul to God as a three-stage journey through the symbolic landscapes of the Old Testament. The purification of the soul from egoistical passions, the enlightenment of the soul by the Holy Spirit and the union with God are compared with the entry into a moonlit desert at night,

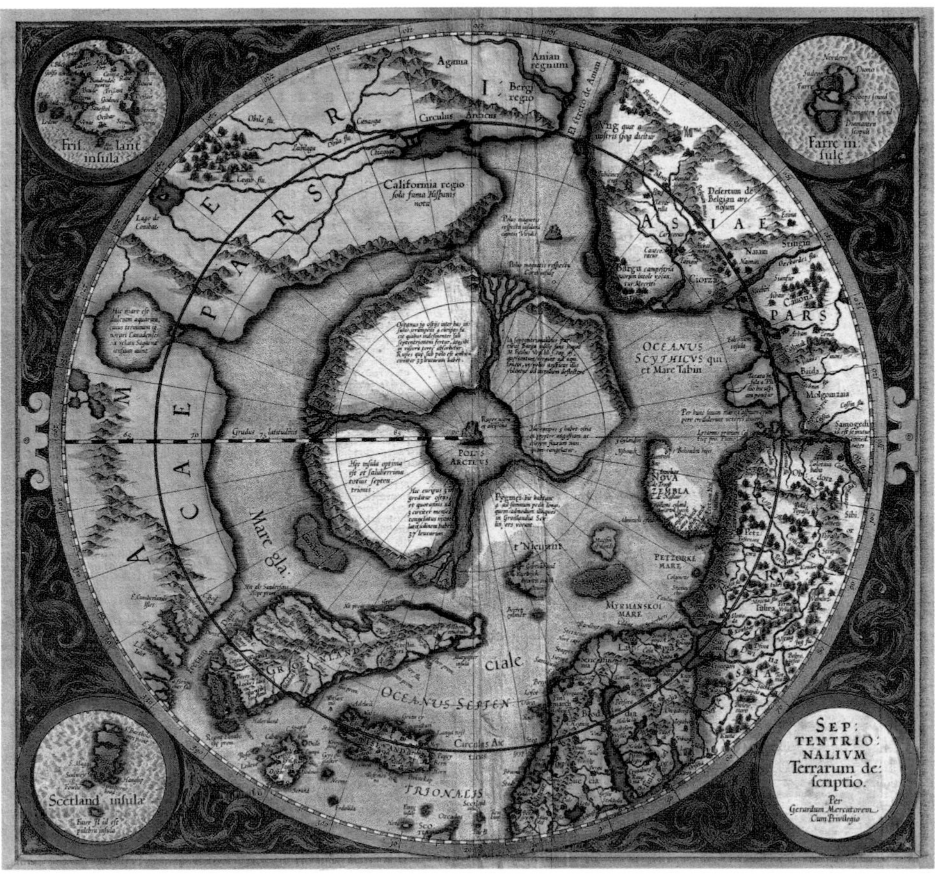

Second edition of Gerardus Mercator's map of the North Pole, featuring the Rupes Nigra at the centre, 1606.

followed by a movement to the seductive, ambivalent fog-covered mountain (Sinai) and, finally, into the impenetrable darkness of a thick cloud and a cleft in the rock on its top.

Knowledge of God is achieved beyond human language, beyond rational understanding, beyond clear vision, beyond landscape. The dark cloud and the cleft on the summit are a precondition to accessing this truth. Visual presence conceals spiritual absence; visual absence invites divine presence. While using landscape imagery, the *Life of Moses* ultimately transcends landscape; it defies the language of cartography – and of mountains. In mapping an ascent, it opens up an infinite abyss. 'Because the one limit of perfection is the fact that it has no limit', asks Gregory, 'how then would [one] arrive at the sought-for boundary when he can find no boundary?'[50]

Mountains and demons

The fear inspired by their dramatic morphology and uncanny atmosphere turned a number of peaks around the world into sites connected to the Devil and hell. Nordic lore populated the misty heights of the Harz Mountains, a forested range in central Germany, with satanic spirits and witches. In the Middle Ages, Hekla, an Icelandic active volcano of 1,491 m, was deemed to be the entrance to the underworld. 'The renowned fiery cauldron of Sicily, which men call Hell's chimney . . . is affirmed to be like a small furnace compared to this enormous inferno', wrote a Cistercian monk in 1180.[51] Four centuries later, a physician from mainland Europe still described how 'out of Hekla's bottomless abyss . . . rise miserable cries and loud wailings, so that these lamentations may be heard for many miles around'.[52] As *axes mundi*, mountains like the Hindu and Buddhist Meru or Dante's Purgatory linked heaven to the mouth of hell and thus conjoined the two extremes of humanity's experience of the sacred in a single cosmographic image.

In the Bible mountains are portrayed as sites of divine revelation as much as demonic sites. The most famous example is the Mount of Temptation, the 'exceedingly high mountain' to whose summit Christ was transported by Satan, after a forty-day fast in the desert at the beginning of his earthly ministry. Here Jesus is tempted with the simultaneous view of 'all the kingdoms of the world and the glory of them' (Matthew 4:8) and he is promised their dominion. Unlike Adam, he resists, thus showing mankind the primacy of spirituality over mundane things.

Unlike other biblical peaks such as Sinai, Tabor or Ararat, the actual location of the Mount of Temptation remains unknown. But where scriptural evidence lacks, tradition intervenes. The Eastern and Roman Catholic churches map the mountain on the Judaean desert and identify it with Jabal al-Qarantal, a 366-m limestone cliff named after Christ's forty-day fast. The spot is marked by a Greek Orthodox monastery overlooking Jericho and the Jordan Valley. In the fourteenth century an Italian pilgrim described Jabal al-Qarantal as 'a very high mountain . . . climbed with much difficulty and danger, for it is higher than all the other mountains'.[53] Later accounts continued to emphasize the prohibitive nature of the cliff. For example, Frank DeHass, the American consul in Jerusalem in the late nineteenth century, wrote of his ascent:

> We had to clamber from rock to rock on our hands and knees, till we gained a shelf at a dizzy height, where we had just room to stand. Here we halted for breath; then, crawling along the brink of the precipice on a narrow ledge, we came to a projecting rock round which it seemed impossible to pass. But others had gone before, and we must follow. Rounding this point was frightful. We shudder to think of that hazardous feat. The path in places was so narrow that

View from Mount of Temptation, overlooking Jericho.

> if a fragment of the rock had given way, or we had lost our balance, or had our feet slipped but an inch, instant death would have followed.[54]

Other local traditions identify Mount of Temptation with more or less dramatic peaks around the Mediterranean: the dark cone of Mount Athos looming on the Aegean; Mount Saitani, a steep stony hill abruptly rising from the bare surrounding plateau of Rovòlakka in Evrytania, central Greece;[55] and Tibidabo, a prominence overlooking Barcelona and deriving its name from the Latin Vulgate Bible verse: 'Et dixit illi haec *tibi* omnia *dabo* si cadens adoraveris me' ('And [Satan] saith unto [Christ]: All these things *will I give thee*, if thou wilt fall down and worship me'). The last is topped by a church surrounded by panoramic terraces and an amusement park – perhaps the ultimate reminder of the futility and deception of mundane pleasures.

The Mount of Temptation is but the climax of a broader scriptural narrative of mountains as negative sites or sites of fear: from the Isaianic association of mountains with pride ('Every valley shall be exalted, and every mountain and hill shall be made low', Isaiah 40:4; cf. Luke 3:5) to the mountain bastions protecting the world from the apocalyptic tribes of Gog and Magog, and, perhaps above all, the continual juxtaposition of the realm of humans, with its cultivated landscapes, to that of craggy wilderness, the abode of evil spirits.

In the Old Testament wilderness is the space of disobedience; it is an unbounded desolate extent made of thorns and thistles to which Adam and Eve were banned after the Fall. Wilderness is also a space for trial and purification, as it continues to be in the New Testament – from the Israelites' forty-year wanderings through the lifeless solitudes of Sinai to Jesus' forty-day fast in the Judaean desert and his repeated retreats to solitary places, including mountaintops.

For ancient Mediterranean people, beautiful nature was essentially domesticated nature: orchards, cultivated fields, vineyards, gardens – productive nature and landscapes imprinted

View of Barcelona from Tibidabo.

with human order. By contrast, mountains were simply inhospitable, desolate, hostile. Such hostility is personified in a Greek stele probably dating back to the third century BC. Here Parnassus features as a frightful giant rising up between two eminences, his wild hair and beard radiating from his face in long locks and tufts.[56] As Lucretius wrote one century thereafter, 'terror and fright fill thickets, mountains and deep forests, terrible places it is almost always in our power to avoid'.[57]

On the southern shore of the Mediterranean of late antiquity, there was, however, a small minority of people who chose such unhomely places as their dwelling precisely for this reason: the Christian hermits. Anthony the Great, the 'father' of desert asceticism, abandoned the fertile Nile Valley for the desolate interior of the labyrinthine 1,200-m-high South Galala Plateau in the north of Egypt's Eastern Desert, also known as the Thebaid region.[58] Having journeyed for three days and three nights, his biographer wrote in the fourth century, Anthony came 'to a very lofty mountain, and at the foot of the mountain ran a clear spring, whose waters were sweet and very cold; outside there was a plain and a few uncared-for palm trees'.[59] The holy man fell in love with the place and established himself there for the rest of his days.

The Euthycles stele featuring a personification of Parnassus, from the Mouseion on Helicon, late 3rd century BC, Athens.

To the occasional visitor, however, the mountain (known in the medieval West also as Mount Climax) was less a place of delight than a *locus horridus*. Such fame grew through the centuries. A fifteenth-century pilgrim, for example, described the mountain as a most dreadful place occupied by 'women notable for long beards' who spent their time 'most cruelly in hunting, have tigers instead of dogs, and breed leopards and lions'.[60] But a challenging environment was precisely what Anthony was after. And he was not alone. Through his example, the holy man, his biographer wrote, 'persuaded many to take up the solitary life . . . From then on, there were monasteries in the mountains, and the desert was made a city by monks, who left their own people and registered themselves for the citizenship in the heavens.'[61]

Anthony and his successors established themselves in the wilderness in direct imitation of Christ. The wilderness was believed to be populated by demons who had been forced to dwell there after they were expelled from the waters when Christ entered Jordan. Anthony and other champions of Christianity went out into the wilderness precisely to battle them, on their own turf.

Anthony's dramatic fights against the demons became a resilient Christian topos. As with Christ's Temptation, they repeatedly feature in Western art and literature. In *The Torment of Saint Anthony* (1487–8), Michelangelo's earliest known painting, the two scenes – Christ's and Anthony's temptations – are conflated. The Egyptian ascetic is captured within a confused mass of monstrous demons and raised high in the air, near the edge of a lofty cliff. Underneath spreads a peaceful verdant landscape resembling more the countryside of central Italy than the Nile Valley. Blue mountain ranges loom on a distant, slightly curving horizon, as if to reinforce the illusion of global mastery offered by the demonic view from above.

The *Life of Anthony*, by St Athanasius, not only shaped literary and artistic imaginations, but ways of life of entire generations of Christian solitaries, especially in the Byzantine East. As Arab invasions forced hermits out of the Egyptian and Palestinian deserts, a new phenomenon spread throughout the Byzantine Empire – that of holy mountains. Between the fifth and eleventh centuries craggy peaks in Syria, Bithynia, Macedonia and other regions of the empire started to be populated by charismatic holy men and their disciples. Their remoteness from the inhabited world, the harsh living conditions dictated by their challenging environment and perhaps also the numinous quality of their rugged landscapes made these

often inaccessible peaks irresistible destinations for solitaries, such as Anthony, who sought spiritual quietness.

Initially Byzantine holy mountains were attributed an aura of holiness because of the presence of charismatic saints. Their fame often attracted disciples and pilgrims, which eventually led to the establishment of organized monastic communities. Even though, unlike Sinai or Tabor, they were not related to biblical events, some of these mountains often ended up surpassing their scriptural counterparts in popularity. Thus in the tenth century Mount Athos came to be known as the 'Holy Mountain' of Orthodoxy, a title that it retains to the present day, along with its twenty Byzantine monastic foundations.[62]

As with Anthony's Thebaid, these wild mountain regions were also haunts of demons, and one of the main tasks of the newly settled hermit was to defeat them and purify these places. When in the early eleventh century St Lazaros settled on the desolate slopes of Mount Galesion in Asia Minor, his biographer informs us, the mountain was infested by evil spirits, 'especially in the cave and in the gorge':

> Once [a monk who was going to see to Lazaros] had already reached the middle of the gorge when the demons suddenly grabbed him by the hair and flung him to the ground; they then began dragging him to a considerable distance . . . Another time when he had been stupefied by the demons, [the same monk] left the monastery of the Saviour and ran into the steep part of the gorge and if the brothers had not run more quickly and grabbed him, he would have certainly flung himself over the edge. Often, too, while the demons were walking about, they would throw stones at him, so that, as a result, he could not go anywhere by himself.[63]

Byzantine mountains like Galesion were trial chambers for the spirit. They were settings for the struggle between good and evil. Stories of temptations and physical battles against demonic creatures crowd the biographies of Byzantine saints of all times, often assuming dramatic tones akin to the Temptation of Saint Anthony.

Michelangelo, *The Torment of Saint Anthony*, 1487–8, oil and tempera on wood.

> One night monk Symeon saw his cave entirely filled with sparkling coals. Straightaway then it seemed to him that some [demons] fell on him with a shout and, having laid hold of him, one of his head and the other his feet, they suddenly hurled him to the ground; and they hit him so hard that he became unconscious from such a beating. After they had beaten him a great deal, they lifted him up in the air and, taking him to the mouth of the cave, suspended him there until the *semantron* of the church struck.[64]

Hermits' eventual victories over demons and their movement to ever higher and more desolate mountain regions marked their upward spiritual trajectories on the model of St John's *Ladder of Virtue*, an ascetic treatise composed in the sixth century in which the reader is taken through thirty virtues (each corresponding to a rung of the ladder). The ascetics who successfully managed to resist the demons of temptation were transported from earth to heaven. By contrast, those who fell victim to the evils of passion were destined for the mouth of hell, as illustrated by manuscript illuminations and icons. Altitude equated beatitude.

Loci horridi

The *Life of Anthony* and the desert myth also had an impact in the Latin West. In the fifth century, for example, as the Marche and other regions in the Italian peninsula that had been influenced by Byzantium grew wilder due to the crisis of the Roman system, local anchorites transformed the Apennines into a 'new Thebaid' – a phenomenon which endured until the seventh century, when it was largely supplanted by the Benedictine model of organized monasticism.[65] This reinforced the ancient opposition between cultivated and uncultivated land, city and wilderness, centre and periphery, human and non-human.

Through the Middle Ages the most austere Latin monks followed their Eastern counterparts in the selection of particularly inhospitable sites for their brotherhoods – think, for example, of the eleventh-century Benedictine Abbey of

Emmanouēl Tzane-Bouniales, *Saint John's Ladder*, 1663.

L
H
H
H
A
E
D
D
C
C

Montserrat, nested on the highest point of the Catalan lowlands, or the Grande Chartreuse in the French Alps north of Grenoble, situated in a deep gorge.

Likewise, in the second half of the twelfth century the Val Ténébreux (literally 'Dark Valley'), a dramatic gorge in south-western France, became a major focus of Catholic pilgrimage. The shrine and monastic complex are still embedded in the spectacular slope known as Rocamadour (from 'rock' and 'Amator', the holy hermit that previously inhabited the place and established the original shrine). The pilgrim is taken to the sanctuary by 216 steps carved in the rock. As with St John's ladder, the mountain, previously a place of darkness and demons, has become a staircase to heaven in the best biblical tradition.

In 1224, two years before his death, St Francis of Assisi selected Mount Verna, an isolated peak of 1,283 m in the centre of the Tuscan Apennines, for a forty-day retreat of fasting and prayer, during which he received the stigmata. Jacopo Ligozzi's illustration of *The Temptation of Saint Francis* (1612) features the Devil tempting Francis on the edge of a steep ravine, like St Anthony in Michelangelo's painting. Here, however, the saint is not promised beautiful, verdant bird's-eye views. The entire scene is dominated instead by the precipitous verticality of the ravine. The presence of the two protagonists is downplayed. Far more frightening than the Devil is the vertiginous precipice towards which Francis is perilously pushed.

The deep gorges and rugged peaks of Western Europe endured through the centuries as demonic *loci horridi*, populated not only by evil spirits but by other wicked creatures. Stories of encounters with dragons abound until as late as the eighteenth century, from the Catalan Pyrenees to the Swiss Alps. Access to dragon-infested mountains, such as Mount Pilatus, a small peak near Lucerne believed to be the burial site of Pontius Pilate, was often forbidden by local authorities for safety reasons. In the early eighteenth century mountain dragons turned from signs of diabolical contamination into objects of scientific enquiry. Notably, Johann Jacob Scheuchzer, a Zurich professor of physics, compiled a catalogue of Alpine specimens. The best

Raffaele Schiaminossi after Jacopo Ligozzi, *The Temptation of Saint Francis*, 1612, etching.

Johann J. Scheuchzer, 'Dragon of Mons Pilatus', in *Itinera per Helvetiae Alpines* (Travels through the Alpine Regions, 1702–11).

ones, he wrote, 'were to be found in the sparsely inhabited cantons of the Grisons: that land is so mountainous and well provided with caves that it would be odd not to find dragons there'.[66]

On contemporary Italian maps, diabolic place names range from modest hills (Monte del Diavolo, near Taranto, 117 m) to more prominent massifs (such as the Pedata del Diavolo in the Tosco-Emilian Apennine, 1,662 m, and Diavolezza in the Upper Engadine, 2,078 m) and perilous or daunting mountain parts, from the various *forcelle del Diavolo* (Devil's saddles) scattered throughout the country to the *grotte del Diavolo* (Devil's caves), *passi del Diavolo* (Devil's passes), *pizzi del Diavolo* (Devil's peaks), *fosse dell'inferno* (hell's crevasses) and *gole dell'infernaccio* ('gorges of the bad hell') in the region of Umbria. Here the concentration of such toponyms appears particularly dense, testifying to the long cultural association of these mountain places with demons and other evil creatures. Near Orvieto, the Culata del Diavolo marks the butt-landing point of Lucifer's fall from paradise – an inelegant anticlimax to Mount Temptation.[67]

Diabolic mountains are not restricted to the map of the Old Continent, though they appear to be a largely European export. 'Devil peaks', for example, recur in former British colonies.

Mount Pilatus today.

They are found in South Africa (adjacent to Table Mountain), in Australia (in the Flinders range) and in Hong Kong. A handful of them features also in Southern California. In the New Continent names such as Devil's Tower (a breathtaking 1,559-m monolith in Wyoming), or Alaska's Devil Mountain's Lakes (the largest volcanic lakes in the world) capture the eerie aspect of these high places, as well as the feelings of early pioneers towards them.

European settlers regarded mountains as part of the savage wilderness they sought to tame. High places represented inhospitable regions filled with unseen enemies; they were perceived as evil obstacles to their progress through the continent. It was not until the American frontier started to gradually disappear that settlers' attitudes towards American mountains became more positive, and, from haunted abodes, high peaks turned into monuments of the nation.[68]

'The sacred is not merely the unknown, but the unknown we regard as ultimately real. It attracts us because it is unknowable.'[69] Mountains embed a strange paradox: they are the most concretely real of all objects, yet they point to heaven, to the invisible, to the unknowable. They are spatially localized and finite, yet they

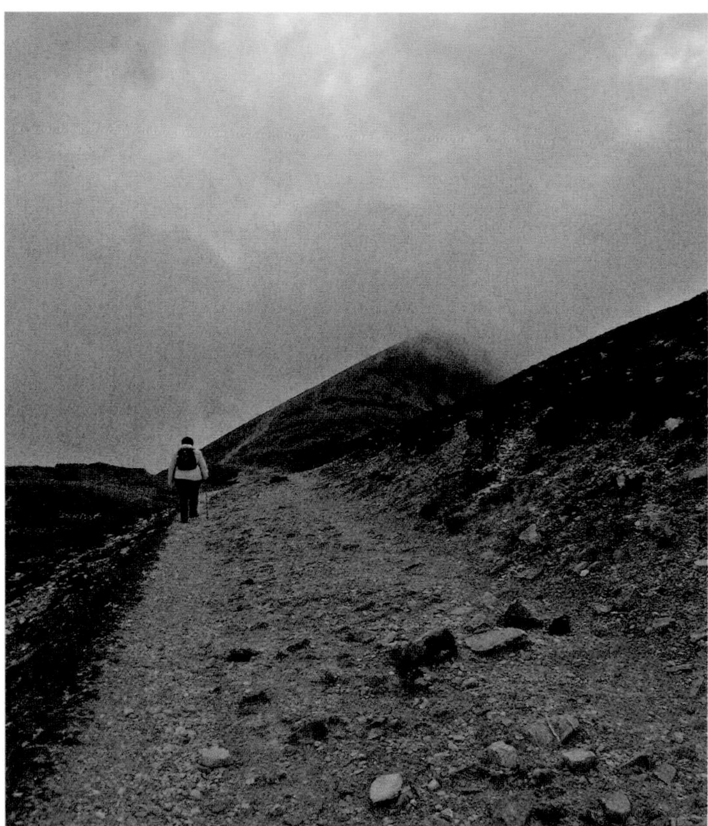

Pilgrimage Pathway, Croagh Patrick, Ireland.

stretch towards the infinite, the eternal, the ungraspable. They lie ambiguously between the material and the abstract, the objective and the subjective, the particular and the universal. The sacred, Mircea Eliade argues, 'reveals absolute reality and at the same time makes orientation possible; hence it founds the world in the sense that it fixes the limits and establishes the order of the world'.[70] As networked landmarks, holy mountains provide meaning and direction in life. As *axes mundi* and peripheral sites of fear, they fix the centre and the limits of our experiential world and of the cosmos; they thus help us feel part of a larger ordered 'whole'. As paradises and liminal or symbolic spaces, they challenge us to chase ever shifting horizons and to move higher and higher, whether physically or spiritually.

Devil's Tower, Wyoming.

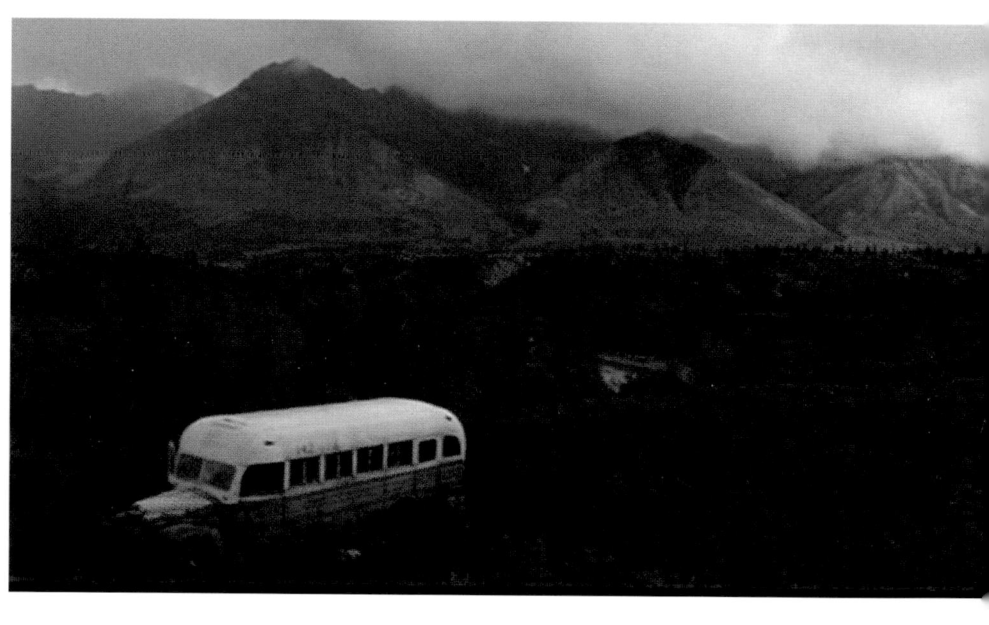

3 Mountains, Life and Death

The old bus where Christopher McCandless spent the last months of his life. Now it is a pilgrimage destination for many.

In 1990 Christopher McCandless, a 22-year-old fresh from finishing his degree and with a promising future ahead of him, decides to cut himself off from civilization. Having destroyed his identification documents, donated all his savings to a charity and changed his name (to Alexander Supertramp), he disappears, leaving his family in anguish and ultimately despair. Two years later his corpse is found, along with his diary, in an abandoned bus in a remote mountain area north of Mount Mackinley in the Denali National Park in Alaska. In 1996 the story is turned into a best-seller by the adventure writer Jon Krakauer and in 2007 it is adapted into the prize-winning movie *Into the Wild*, under the direction of Sean Penn.[1]

During the two years following his disappearance, McCandless lives the life of a vagabond, traversing the States from East to West, kayaking through the Grand Canyon into Mexico and then heading north to Alaska, his ultimate destination. At the root of his escape into the wild is an uneasiness with his complicated family situation and with society at large, which he perceives as being dominated by a hollow materialism. More profoundly, however, there is also an insatiable thirst for meaning, a continuous desire to explore and redefine his own limits through solitude and intimate contact with nature. McCandless's movement through space is but the physical manifestation of a parallel, endless inner journey. 'The joy of life comes from our encounters with new experiences, and hence there is no greater joy than to have an endlessly changing horizon, for each day to have a new and different sun,' comments Krakauer.[2] Looming on the horizon of McCandless's imagination are the majestic and desolate snow-capped peaks of Alaska.

Once reached, however, the wild mountains end up becoming his grave. Having settled into the abandoned bus, the young man manages to sustain himself for four months, but eventually starves to death. Having realized that wilderness is cruel and uncaring, McCandless also eventually realizes that 'happiness [is] only real when shared'. He attempts to head back, but the river he crossed in winter has now become impossible to traverse due to the thaw. Slowly dying, he continues to record the final phase of his journey of self-realization and silently submits to his fate, as he imagines his family for one last time. 'What if I were smiling and running into your arms? Would you see then what I see now?'

Emblem of ultimate freedom, idyllic setting and mortal trap, the Alaskan mountains are also the site of important revelations, whereby life eventually acquires a new meaning and McCandless's soul is reconciled, if only close to the point of death. His story can be read as the ultimate pilgrimage in a Western secularized society where contact with the transcendent is sought more through the immersion in extreme environments than, say, in a church. McCandless's, however, is also but one of innumerable stories in which life and death intersect on and through mountains. For mountains are meeting points of extremes.

As the epitomes of wild places, mountains possess a singular capacity to simultaneously allure and threaten, to reveal the meaning of life and to take life away. Their blue silhouettes looming on distant horizons, the pristine beauty of the landscape, the deceitful intimacy of the rock, the silence of the abyss, the pain and pleasure of the ascent, the gaze over the clouds, the occasional encounter with some charismatic animal, continue to draw thousands of individuals to unforgiving summits. Every year, mountains' perverse charm causes dozens of violent deaths. Mont Blanc alone has killed over 1,000 people since it was first climbed; the Matterhorn 500; Everest over 200; and the list continues. Remote mountain passes are also frequent theatres

Party ascending Everest's 'death zone'.

of armed conflicts, deathly barriers and corridors in which every day dozens of anonymous soldiers and refugees lose their lives as they hope or search for a better life. This chapter explores encounters between life and death in the mountains.

Life at the extreme

Mountains fall into what the early twentieth-century geographer H. J. Fleure named 'regions of difficulty', that is, areas of scant resources for dense human occupation that 'refuse sensible increment even to prolonged effort'.[3] Low temperatures, regular snowfall, high winds, rarefied air and a rough topography unsuitable for agriculture make high places difficult, sometimes even impossible, to inhabit. The highest-known permanent settlement in the world is at 5,100 m, in the Andean mining village of La Rinconada in southern Peru.[4] But this is an exception. Only a tiny fraction of the world's population dwells above an altitude of 2,500 m and an even tinier fraction (less than 0.3 per cent) above 3,000 m. Certain mountain regions are simply off limits to humans. Above 8,000 m (for example, on the summits of Everest and K2), there is insufficient oxygen to support human life. This is known as the 'death zone'. 'No animal or plant could exist here . . . in this absence of all life, this utter destitution of all nature', wrote the French mountaineer Maurice Herzog of Annapurna, one of the world's fourteen peaks above 8,000 m.[5]

Even at lower altitudes, physical discomfort, isolation and danger make mountains transitory places for human beings, but other creatures have nevertheless made them their permanent habitat. Perennial grasses, forbs, cushion plants, mosses and lichens are recurring presences on high summits around the world, now decorating their upper steeps in fanciful patterns, now erupting unexpectedly from the most improbable fissures in the rock. Faced with low temperature extremes, strong winds and increased ultraviolet radiation, these alpine plants have developed various techniques of adaptation. Waxed surfaces, for example, filter tissue damage and light radiation; blisters and

The Himalayan yak (*Bos grunniens*).

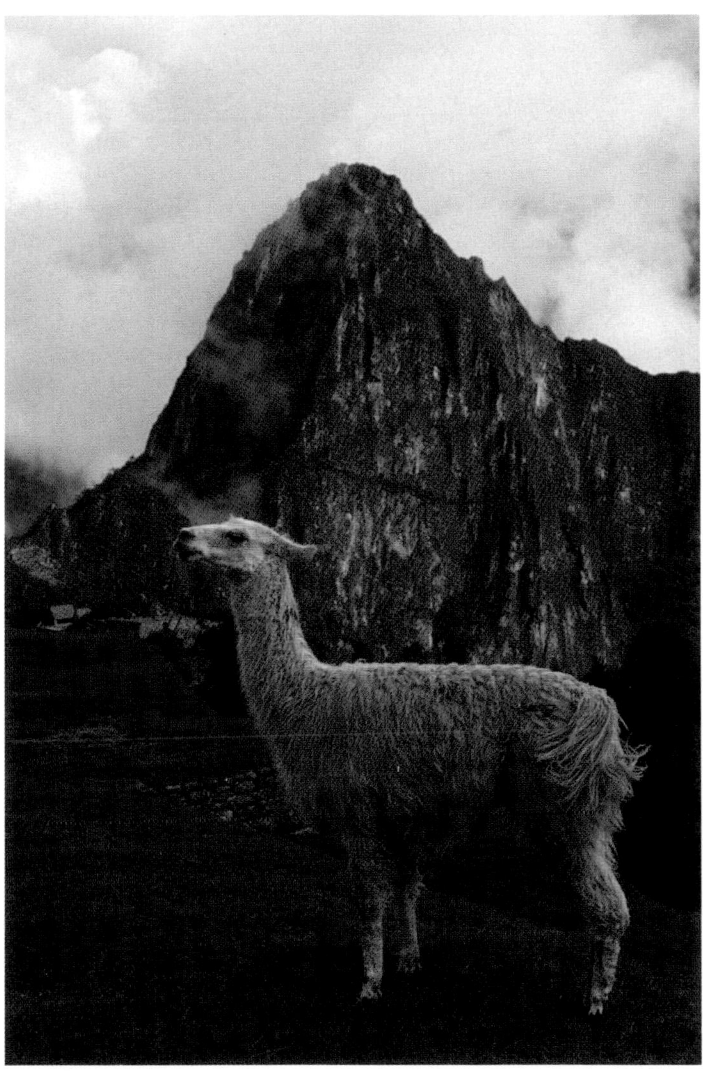

Llama at Machu Picchu (*Lama glama*).

hair keep wind and air away from the plants' surfaces; deep roots help them prevent water loss.

Certain animals have likewise developed special adaptations for living at extreme altitudes: enhanced lung capacity and haemoglobin levels to allow them to cope with oxygen deficiency, thicker coats than other species to face the cold, the ability to slow the heart rate and survive with reduced blood flow (preventing the loss of excess heat), and the rapid maturation of cubs to ensure self-sufficiency before the arrival of the winter months.

In the Himalaya, the yak (*Bos grunniens*) is adapted to live at over 3,000 m and suffers heat stress below this. Its smaller domesticated counterpart is used by the Sherpa people as a beast of burden across high mountain passes. The zomo, a cross between a yak and a cow, is able to live at altitudes of 2,100–3,660 m. The fat-tailed Himalayan sheep can store water in its tail and thus survive arid conditions. In the Andes indigenous guanacos, llamas and alpacas are adapted to survive the diurnal extremes of temperature of the high puna and páramo grasslands. Likewise,

Machu Picchu, Peru.

mountain goats are found in North American mountains at altitudes of up to 3,000 m, whereas their European relatives the angora chamois and bouquetin dwell in the Alps between 1,000 and 3,000 m, spending the summer on the high pastures and glaciers and migrating to the forests in winter.[6]

As a rule, organic diversity and life itself decrease as one climbs up towards the summit. Coniferous trees give way to shrubs, shrubs to tundra, tundra to bare jagged rock and snow. In the highest regions of the Swiss Alps, for example, one finds no more than eight species of birds, in contrast to the 27 in the shrub zone underneath and the 96 in the coniferous forest.

While ancient civilizations like the Minoans and Inca built their settlements and shrines in high places – think of Machu Picchu at 2,430 m – the first crosses were implanted on Alpine summits only in the nineteenth century. Unlike the long litany of Alpine villages and passes bearing saints' names – St Moritz, St Gallen, San Vigilio di Marebbe or San Bernardino, for example – no mountain peak was entitled to a holy protector. Summits remained for the most part unnamed and untrodden.[7] Life was in the valleys; death, up in the mountains.

Heights and depths

When did mountains start to exercise their spell in the West? When and why did they begin to be longed for rather than feared?

Until the seventeenth century Swiss and Italian high places were considered ugly and dangerous regions to be carefully avoided; 150 years later they had become main tourist attractions and among the most spiritually uplifting places on the planet.[8] Unlike the holy men of the past, the new brave men who ventured to mountain regions were not necessarily religious. They did not set out to encounter God or fight demons; theirs was rather 'an experiment in sensation' – in the sublime.

The first travellers on the Grand Tour stumbled into the Alps almost accidentally, on their way to Rome, Naples and other Italian centres identified with classical culture. And yet so dramatic was the impact that such high places ended up

Salvator Rosa, *Rocky Landscape with a Huntsman and Warriors*, 1670, oil on canvas.

becoming attractions in themselves, ones surpassing any man-made works of antiquity. Unlike classical monuments, however, these strange landscapes of rock and snow were not beautiful, but, as the French traveller and poet Charles-Julien Lioult de Chênedollé put it, 'beautiful and horrible', like the haunting landscapes painted by Salvator Rosa.[9] Their vertiginous depths and craggy heights displaced the viewer; they caused what Kant would later call a 'negative pleasure', an uncanny mix of attraction and repulsion – 'a delightful Horror, a terrible Joy', John Dennis wrote in 1693. 'We walk'd upon the very brink, in a literal sense, of Destruction; one Stumble, and both Life and Carcass had been at once destroy'd.'[10]

Alpine Grand Tours were designed precisely to take their protagonists to the edge, like St Francis in Ligozzi's drawing – this time to flirt with, rather than resist, diabolic temptation. 'A giddy path follows the edge of the precipice', wrote Friedrich Schiller on the Gotthard:

> You walk between life and death. Two threatening peaks shut in the solitary road. Traverse noiselessly this place of terror; fear to awaken the sleeping avalanche. The bridge which crosses the frightful abyss, no man would have dared

> to build. Below, without power to shake it, growls and foams the torrent. A sombre arch seems to conduct to the empire of dead.[11]

According to the theologian Edward Burke, it was 'in shadow and darkness, in dread and trembling, in caves and chasms, at the edge of the precipice' that the sublime was to be discovered.[12] The vertiginous disorientation produced by these forms, with their dramatic chiaroscuro, misty heights and gloomy abysses, was best captured by the work of eighteenth-century painters. John Robert Cozens (1752–1797) went as deep as to penetrate the earth 'as if sucked through some Virgilian vortex at the mouth of hell'.[13]

Later Romantics took the Alps as metaphors of infinity; they found a secret correspondence between their deep gorges and the profundities of the human soul. 'One cannot stand on any lofty mountain summit without experiencing feelings that move the heart to its profoundest depths', Reverend Joel Tyler Headley wrote in the preface to his popular *Mountain Adventures* (1876).[14] To be profound was to plunge into such dark abysses and, in an imaginary flight, soar up to impossible summits, towards boundless heavens. As he contemplated the snow-capped summit of Mont Blanc from the valley of Chamonix, Shelley wrote:

> The everlasting universe of things
> Flows through the mind, and rolls its rapid waves,
> Now dark – now glittering – now reflecting gloom –
> Now lending splendour, where from secret springs
> The source of human thought its tribute brings.[15]

Through Wordsworth's verses, Alpine drama was transported to the Lake District. The reader is walked to Snowdon under the moonlight, is offered views over misty cragginesses and dark abysses and summoned up into the mountains where 'he had been alone / Amid the heart of many thousand mists, / That came to him, and left him, on the heights'.[16] Sometimes lonely characters emerge from the depths of foggy, desolate

John Robert Cozens, *Satan Summoning his Legions*, *c.* 1776, watercolour on paper.

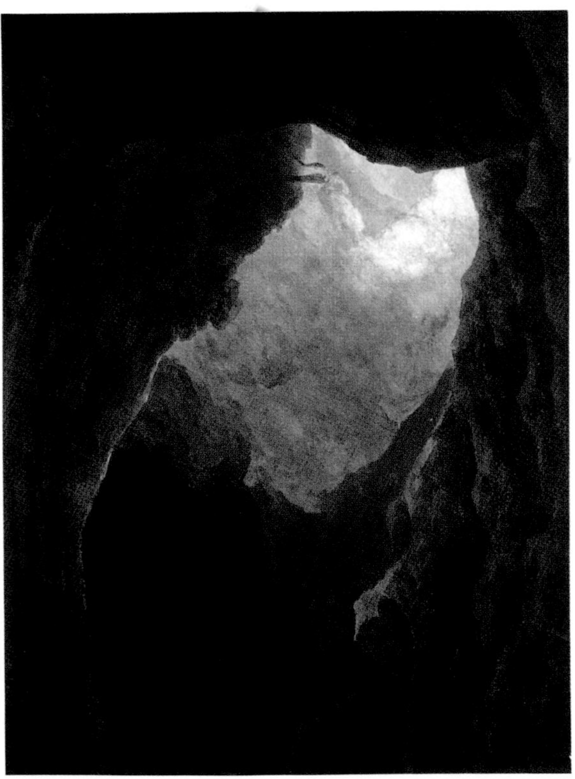

John Robert Cozens, *Cavern in the Campagna*, 1786, watercolour.

rocky landscapes. At other times they mysteriously disappear into them. At others, they simply merge with them.[17] The old 'Leech-gatherer' in his *Lyrical Ballads*, for example, features 'as a huge stone . . . sometimes seen to lie / Couched on the bald top of an eminence'. In 'The Excursion', the Solitary, a character who, afflicted by the death of his wife and children and disenchanted with the French Revolution, has chosen to withdraw from society, is revealed a glimpse of heaven through the fog. As peaks and clouds merge in a vision of the celestial city,

> my heart
> Swelled in my breast – I have been dead, I cried,
> 'And now I live! Oh wherefore do I live?'
> And with that pang I prayed to be no more.

The modern sublime has been interpreted as a response to the displacement caused by the Copernican revolution and the *furor astronomicus* that followed Galileo's invention of the telescope. As the reassuringly self-enclosed space of the ancient Aristotelian cosmos opened up, Western Europeans were confronted with infinity. Humans lost their central place in the universe to discover themselves as insignificant dwellers of a planet floating in unbounded black spaces. Creation was no longer confronted through prayer, but agnostically. 'The eternal silence of these infinite spaces terrifies me', famously wrote Pascal.

At the same time, measuring oneself against the forces of nature strengthened self-esteem and boasted self-confidence – and thus ultimately created a new, hubristic form of anthropocentrism. The most dramatic of all sceneries, mountains took modern subjects away from the everyday and forced them into introspection – 'to reveal something of themselves'. Such feelings animated mountaineering, as they continue to do today. 'Why do we climb mountains? Why do we risk our lives in the high places of the world?', asked the American writer, broadcaster and mountaineer Lowell Thomas in the late 1960s. 'Certainly not just for the view. Is it to meet the challenge of a so-called impossible peak simply because it is there? Or is it perhaps to meet a greater challenge within man himself?'[18]

Death in the mountains

Mountaineering as a sport flourished in the nineteenth century. Mont Blanc, whose peaks were collectively known in French as the *montagne maudite* (literally, the Accursed Mountain), was first ascended in 1786; the infamous Eiger in 1858. The Matterhorn only in 1865. The first Alpine Club was established in London in 1857; its Austrian, Italian, Swiss and German counterparts followed in the 1860s. By the end of the century all the Alps had been climbed and almost all the Alpine passes mapped.

Mountains became popular tourist destinations and temporary escapes from increasingly industrialized urban areas, what Leslie Stephen called 'the playground of Europe', an arena of 'recreation, but also one of re-creation'.[19] As the British Empire reached further domestic stability and prosperity, Victorians grew increasingly fond of risk-taking. 'The middle classes needed a danger valve', Robert Macfarlane comments, 'somewhere they could let off the steam which built up through cosseted urban living – and the Alps were just the place, for there everyone could find their own levels of risk'.[20]

Regardless of their frequency, violent deaths, or, rather, the possibility of a violent death, contributed to the lure of mountaineering in the nineteenth-century Western imagination. From abodes of dragons and demons, as they had been for centuries, certain Alpine peaks became cursed with the memory of human deaths – and the sight of death was essential to Europeans' mountain experience. Mountains were to become funerary monuments, as Shelley prophetically wrote in 'Mont Blanc' (1816), or, more simply, places of encounter between life and death.

Mountain death was a popular topic in Victorian travel literature. Headley, for example, opened his *Mountain Adventures* with a long list of Alpine tragedies. The most dramatic, he wrote, had recently occurred on Mont Blanc, as two Americans, a Scottish clergyman and eight guides were suddenly caught in a blizzard. The unfortunate party was last seen by tourists looking through their telescopes at Chamonix, as they were striving to survive the tempest. Then the snow hid them from sight and they were never seen alive again.

> At length some black spots could be detected on the white surface, and 50 men set out toward them. They proved to be five bodies of the unhappy party. One of these was Dr Bean of Baltimore, who was found sitting up in the snow with his head fallen forward on his hands. For two days he had sat there nearly three miles in the heavens, listening to the roaring of the storm that shut out the world below him. During these two dreadful days, he traced with his stiffening fingers a few lines in a diary which was found upon him . . . [The last entry, addressed to his wife, read:] 'I die believing in Jesus Christ, with the sweet thought of my family, my friendships and all. I hope we shall meet in heaven. Yours always.'[21]

Stories of mountain tragedies all resemble one another: the fatal attraction and 'possession' exercised by the mountain, its unfeeling and unforgiving nature, climbers' failed heroic feats, immortal glory buried by the snow and the wind. Out of the interminable list of deathly accidents, however, two not only stand out in the history of mountaineering but have become icons in Western popular culture.

The first occurred on the Matterhorn in 1865. For nearly a century since the conquest of Mont Blanc, this peak remained mountaineers' unattainable object of desire. While lower than Mont Blanc, the Matterhorn is far more difficult to climb – and therein lay its spell. Its sharp sword-like summit, steep cliffs and glaciers sliding underfoot into horrid abysses made it known in the Alps as 'the awful mountain'. Well into the nineteenth century it was a common belief that the Devil dwelled near its top and hurled rocks down into the valley.[22]

Edward Whymper, a young British artist who had initially come to the Alps to sketch, was one of those who remained 'possessed' by the Matterhorn and resolved to climb it. After three failed attempts, he eventually reached the summit with a party of seven, including Michel Croz, one of the most famed mountaineers of the time. During the descent, one of them suddenly slipped. As the seven were roped together, his forward motion sent him into Croz, who in turn dragged down two other party members. 'All this was the work of a moment', Whymper records

> Old Peter and I planted ourselves as firmly as the rocks would permit: the rope was taut between us, and the jerk came unto us both as one man. We held; but the rope broke midway between Taugwalder and Lord Francis Douglas. For a few moments we saw our unfortunate companions sliding downwards on their backs, and spreading out their hands, endeavouring to save themselves . . . They disappeared one by one and fell from precipice to precipice on the Matterhorngletscher below, a distance of nearly 4,000 feet in height . . . For the space of half an hour we remained on the spot without moving a single step.[23]

That moment haunted Whymper for the rest of his life. Solitary and silent, he roamed the world in search of new peaks to conquer. And yet,

> Every night, do you understand, I see my comrades of the Matterhorn slipping on their backs, their arms outstretched, one after the other, in perfect order at equal distances – Croz the guide, first, then Hadow, then Hudson and lastly Douglas. Yes, I shall always see them.[24]

The accident inspired numerous movies and documentaries, including Mario Bonnard and Nunzio Malasomma's silent film *Struggle for the Matterhorn* (1928), which was later remade into *The Challenge* by Milton Rosmer and Luis Trenker (1938). Its iconic status, however, owes much to Gustave Doré's dramatic drawing of the Fall. As opposed to Romantic paintings of the Alps, where tiny human bodies (when present) are almost lost

Original 1928 poster for the silent film *Struggle for the Matterhorn.*

in the magnitude of overwhelming sceneries, here they are tragically brought to the fore. 'One feels the agony of the moment as he sees the four men roped together frozen in mid-air, as they plunge wildly through space from that lofty precipice', in a fashion redolent of Salvator Rosa.[25]

The second most well-known tragedy occurred in 1924, this time far away from Europe, on the slopes of Everest. If the Matterhorn looms in the Western imagination as the icon of difficulty, the 'roof of the world' has become a symbol of the most extreme human aspirations and achievements. As the

Gustave Doré, *The Fatal Accident on the First Ascent of the Matterhorn*, 1865, lithograph.

ultimate junction between earth and heaven, Everest is what Macfarlane called 'the greatest mountain of the mind'.[26] Its peak was conquered in 1953 by the New Zealand mountaineer Edmund Hillary and the Nepali Tenzing Norgay at the height of the Cold War. However, a number of other attempts had been made in previous decades, among them the ill-fated expedition of 1924 by the Englishman George Leigh Mallory.

Given the altitude of the mountain, the expedition was planned in multiple stages. Various camps were established at different elevations and used as points of departure. The last, camp VI, was set as near to the summit as possible, so as to leave no more than 600 m for the final dash to the summit. This would be attempted in waves of two-man teams. The first two attempts were unsuccessful. Adverse weather, physical exhaustion and illness forced both pairs back to the camp. Mallory, however, was resolved to reach the summit and thus made a third attempt. As the wind dropped and the sun came out, he and his partner Irvine set off and managed to reach the higher slopes of the mountain.

On the third day Noel Odell, the expedition's geologist, walked away from the camp and reached the top of a small cliff jutting out of the ridge. 'As I reached the top', he later wrote,

> There was a sudden clearing of the atmosphere above me, and I saw the whole summit ridge and final peak of Everest unveiled. I noticed far away on a snow slope leading up to what seemed to me to be the last step but one from the base of the final pyramid, a tiny object moving and approaching the rock step. A second object followed, and then the first climbed to the top of the step. As I stood intently watching this dramatic appearance, the scene became enveloped in cloud.[27]

Last known photo of Mallory and Irvine.

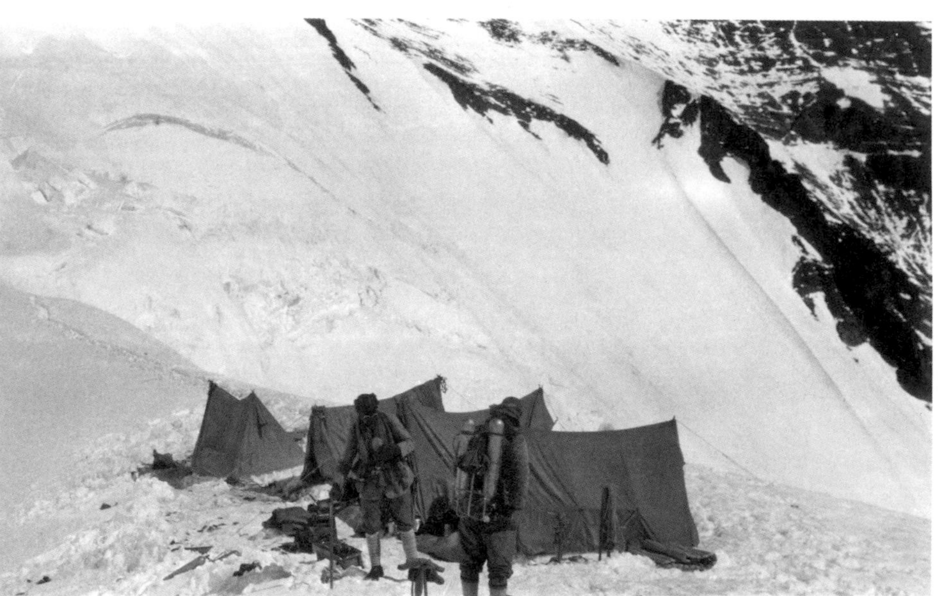

This passage, one of the most famous and evocative in the history of mountaineering, became an icon of the ill-fated expedition: the 'tiny spots' were lost in the grandeur of the mountain backdrop, symbolizing the magnitude of Mallory and Irvine's task; the cloud behind which they disappeared shrouded the mystery of their fate. That was to be the last time the two men were seen.

As Odell returned to the camp, snow began to fall and the wind to rise. A blizzard swiped the face of the mountain. Thinking that his two companions would have turned back and would find it difficult to find the camp in the mist, Odell went out once again, scrambled up a couple of hundred feet and started to shout to direct them. But the two were not near enough to hear him.[28] 'The blizzard ended, but the night fell, a night of frantic looking for the light of a flare in the sky. The darkness remained, implacable, terrifying.'[29]

Mountaineering, empire and exploration

In the decades between the ascents of the Matterhorn and Everest, a series of other mighty peaks emerged on the horizons of the Western imagination. In the 1890s, once all the summits in the Alps had been conquered, mountaineering became

Lenana was the Chief Medicine Man of the Maasai, *c.* 1890. Point Lenana on Mount Kenya was named after him by Halford Mackinder.

incorporated into a new rhetoric of imperial exploration and adventure. Before definitively shifting to the Himalaya, British attention moved to the great mountains of Africa.[30] In the accounts of the first European explorers of these peaks, tropical diseases, unruly natives, impenetrable vegetation and exotic ferocious animals made the thin boundary between life and death appear even feebler.

Mount Kenya, at 5,199 m the second highest peak in the continent after Kilimanjaro, was gained in 1899 by Sir Halford Mackinder, the first reader of geography at the University of Oxford. Four years earlier the anthropologist Mary Kingsley had ascended Mount Cameroon (4,040 m), locally called Mungo Mah (The Throne of Thunder), the highest summit in West Africa.[31] 'It is none of my business to go up mountains', she confessed.

> Nevertheless, I feel quite sure that no white man has ever looked on the great peak of Cameroon without a desire arising in his mind to ascend it and know in detail the highest point on the western side of the continent, and indeed one of the highest points in all Africa.[32]

The peaks of the 'dark continent' became less sublime landscapes demanding romantic reverence than summits to be conquered and collected as trophies by stout white men. Rather than platforms from which to humbly contemplate the infinity of heaven and the inscrutable profundities of the human psyche, African peaks were panoramic vantage points from which to master the world (and people) underneath. By violating those summits sacred to the natives, European scientists, explorers and missionaries were determined to demonstrate the superiority of their own creed and of Western rationality over local traditions and superstition.

European 'assaults' of African mountains were inevitably enacted as adventurous displays of imperial power – and yet in most cases they continued to intersect with death, though in different ways from Alpine climbs. Mackinder's expedition included 66 Swahilis, two Maasai guides and 96 Kikuyu. On their way to the top, the party passed through a country devastated by plague and famine. When they reached the base camp, they could not find any food and two of their party were killed by local people. 'The whole caravan was very much upset, and repeatedly asked to go back, until, losing my patience, I set about those nearest to me with my fists. The effect was really marvellous.'[33] By contrast, Kingsley congratulated herself on having never used her gun against her men and having got them 'up so high and back again, undamaged'.[34]

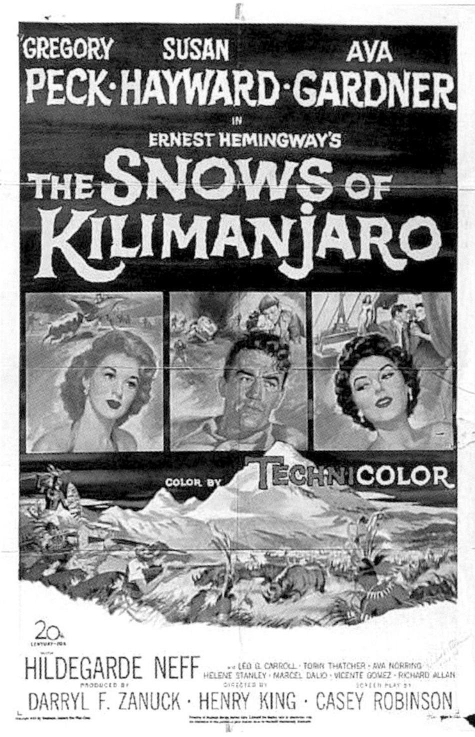

Poster of the film version of *The Snows of Kilimanjaro* (1952).

Heroic mountain feats were not an exclusively British domain. Mount Stanley in the Rwenzori range (Uganda), at 5,091 m the third highest peak in the continent, was successfully ascended for the first time in 1906 by the Italian prince Luigi Amedeo, Duke of Abruzzi, and his party, which included over 300 native porters. Several attempts had been made by previous explorers, but they were always turned back by the thick vegetation, adverse weather or disease. The spectre of death silently followed the duke's party throughout the ascent. The chronicler of the expedition described the scenery as quirky and otherworldly, if not macabre, with 'tall columnar stalks of *labelia* [rising from the ground] like funeral torches, beside huge branching groups of the monster *senecio*'. The spectacle, he relates, was 'too unlike the familiar images, and upon the whole brooded the same grave deadly silence'.[35]

The first European to ascend Kilimanjaro was the German geologist and explorer Hans Meyer in 1889. It was, however, not so much this ascent that captured the Western popular imagination but rather one in 1926, during which another

German, Richard Reusch, discovered the frozen carcass of a leopard near the highest summit. The event inspired Ernest Hemingway's short (and partly autobiographical) story 'The Snows of Kilimanjaro' (1936), making the mountain one of the iconic peaks of modern Western literature and cinema.

'No one has explained what the leopard was seeking at that altitude', the story begins. The answer to the enigma comes only at the end. The protagonist, an American writer named Harry, lies in his tent in the savannah awaiting his slow death of gangrene. Like an old frontiersman (and Hemingway himself), Harry had travelled to Africa seeking a temporary escape from civilization and its discontents. He had hoped that wilderness would restore his life and offer an antidote to his indolence and writer's block. His nearing end causes him to look back at the past and inside himself. Vivid flashbacks from his decadent and over-modernized life in different European cities intertwine with the realization of how little he has achieved in his life and writing. In his final delirium he has a vision of a small plane coming to rescue him. He and the pilot take off and fly into a storm. When they come out of it, Harry realizes where he is being taken: 'great, wide, high, unbelievably white in the sun . . . the square top of Kilimanjaro', the ultimate destination. For the writer, as well as for the leopard, 'the evocation of a sacred, mythical wilderness outside the American historical context comes at a price: death.'[36]

Mountain conflicts

With the end of colonial rule, the great African peaks turned from imperial trophies and sites of death into symbols of rebirth and life. At the moment of liberation in the early 1960s, Kilimanjaro and Mount Kenya, for example, were ascended by local climbers and flares were lit on their summits, announcing the birth of the new nations to the world and radiating a message of hope to the entire continent.[37] By contrast, other mountains and ranges in former and current colonies around the world have been, and continue to be, theatres of some of the

One of the many media images from the war in Afghanistan featuring a mountain landscape in the background.

most dramatic conflicts – from the Golan Heights in the 1973 Arab–Israeli War to the bleak Antarctic mountains of South Georgia during the Falklands War in the early 1980s and the mountains of the Caucasus in the more recent wars in Chechnya. Likewise, barren brown peaks have become the backdrop to the ongoing conflict in the eastern mountain province of Afghanistan following the 2001 U.S. invasion of the country. Unlike Western mountain climbers and explorers, the names of the victims of these conflicts remain largely forgotten or unknown.

Mountain combats are among the most dangerous, as they involve surviving not only the enemy but the extreme weather and treacherous terrain. Strong gusts of wind, lightning, falling rocks, avalanches, crevasses, blizzards, extreme cold and, in some cases, lack of oxygen are all potential causes of death. In this respect the most extreme of all postcolonial mountain disputes is the ongoing conflict between India and Pakistan over the

Siachen Glacier in the Kashmir region, where 5,000 troops from India are permanently stationed.

At the time of partition in 1947 this part of the border between the two countries was left unmapped because the cartographer deemed definition to be unnecessary; the terrain was so unwelcoming and the details so sketchy that none of those responsible for determining the border imagined that the area would become a matter of contention.[38] Since Partition, however, continual skirmishes between the two countries have taken place in this region, which contains some of the highest mountains in the world. Between 1984 and 2012 the Indian Army lost nearly 900 soldiers and the Pakistan Army 1,800, with many more wounded. Of those killed, 97 per cent were killed by the mountain conditions and only 3 per cent by enemy fire.

The Siachen Glacier begins at 5,425 m, in the Saltoro Range of the Himalaya, and it climbs up to 7,773 m, which is a mere 1,066 m less than Mount Everest. Most of the troops are stationed in the valleys below, about 80 km from the road-head, and have to be maintained entirely by air. The soldiers take turns serving at the forward posts, situated at over 6,000 m on the glacier. One in two will die. What mountaineers around the world undertake as a one-off enterprise, the average soldier here does as a matter of duty on a routine posting, living from three to six months at icy heights with temperatures dipping as low as −40°c, an extremely rarefied atmosphere and the continuous threat of

Pakistan stamp issued in 1960 featuring the contested area of Kashmir as 'not yet determined'.

avalanches.[39] On 7 April 2012 an avalanche hit a Pakistani military camp 30 km west of the Siachen Glacier terminus; 129 soldiers and eleven civilians were killed. The Siachen Glacier marks not only a contested boundary between two countries, but the boundary between life and death:

> Last fortnight, we were witness to the funeral of the latest victim, Chandra Bhan, 28, a sepoy of the Rajput Regiment stationed at the forward post of Pahalwan (20,000 ft) which often comes under concentrated Pakistani artillery fire. Bhan, however, did not die because of enemy action. The cause of death was pulmonary embolism – a blood clot in the lungs caused by the rarefied air . . . Bhan had actually died 18 days earlier. His body was laboriously carried down the steep snow-covered slope to Zulu (19,500 ft), the only post in the area with a helipad . . . Bhan's widow – expecting the third child – will get only an urn of her husband's ashes and his ribbons, including a grey and white one awarded to all Siachen veterans. She is fortunate. Many others won't even get that. The bodies of their menfolk – who were trapped in crevasses, buried under avalanches, or lost in blizzards – have never even been recovered.[40]

Because of their strategic importance as natural borders, mountain ranges have a long history as sites of conflict and tragedy. As early as 1812 the Prussian general and military strategist Carl von Clausewitz dedicated an entire chapter of his famous treatise *Principles of War* to mountain combat and the defensive role of high places.[41] Towards the end of the century special mountain troops were established in many European armies. The first elite Alpine infantry unit was formed in 1872 to protect Italy's northern mountain border. During the First World

Dolomites at war at 3,000-m altitude, *La Domenica del Corriere*, 28 January–4 February 1917.

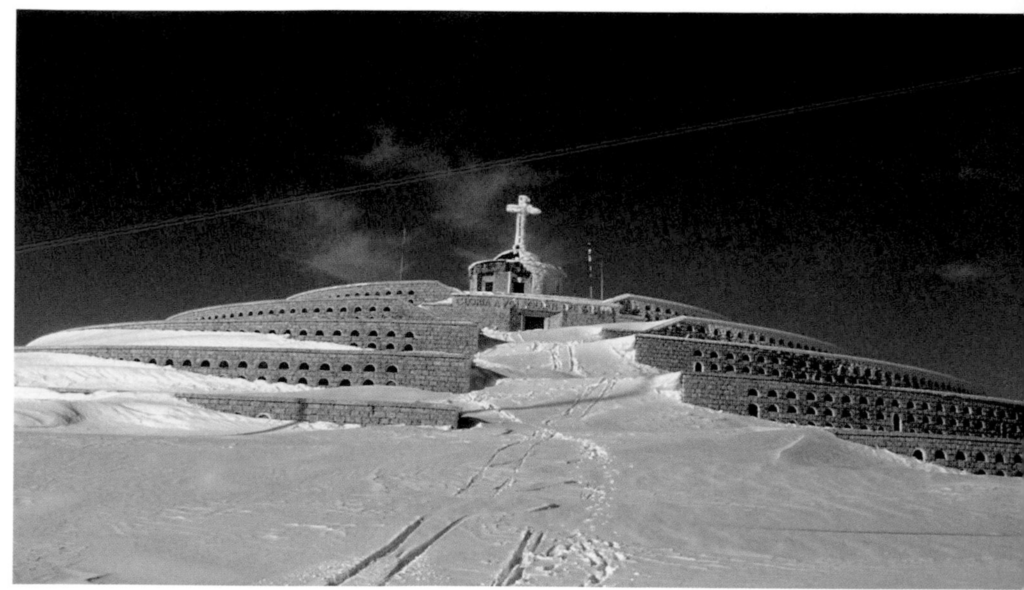

Monte Grappa Ossuary covered by the snow.

War the region became the theatre of the cruellest fighting. Around a million men died in what has been called 'the war in snow and ice'. Most of them were killed by frostbite and avalanches. On the high Alpine sectors of the front, the soldiers lived and fought in year-round whiteness – thick white snow in winter and blinding white limestone in summer. A strip of no man's land separated the Italian and Austro-Hungarian armies:

> Peering at a field cap bobbing above the enemy trench, an Italian soldier reflected on the conditions that made the carnage possible: 'We kill each other like this, coldly, because whatever does touch the sphere of our own life does not exist . . . If I knew anything about that poor lad, if I could once hear him speak, if I could read the letters he carries in his breast, only then would killing him like this seem to be a crime.'[42]

From time to time, remnants of nameless soldiers continue to be retrieved amid the Dolomites' rocks by the occasional climber or excursionist. Others are buried in officially designated memorial landscapes. The most spectacular and disturbing is the military ossuary of Monte Grappa. Here the remnants of over 12,000 mostly anonymous Italian and Austro-Hungarian soldiers silently rest side by side. The monument was designed by the Italian architect Giovanni Greppi in 1935. Located on the top of the mountain, it is formed by five concentric pyramidal circles culminating in a sanctuary dedicated to the Virgin of Mount Grappa.

Yet the whole mountain is a monument. Its slopes still bear the physical scars of the conflict: trenches half hidden by the snow, vast grenade craters filled with Erica flowers, invisible tunnels and caverns piercing into the guts of the mountain, where armaments were once stored. On clear days the site offers its visitors a panorama of the entire front – a 600-km line of fire from Hermada to Adamello. No other spot provides such a comprehensive vista. Such was the view that galvanized the Austrians. And such was the view that transformed the Italians, as for the first time they were enabled to grasp the fate of their country in utter lucidity.[43]

When during the Second World War conflict shifted from the land to the air, mountains assumed special importance as landmarks for pilots. Hence the people of Japan felt betrayed by Mount Fuji, as their very own holy mountain and national symbol suggested to the Americans where to drop the atomic bomb.[44]

Throughout the war and in its aftermath, the mountains of Europe became bulwarks of resistance, as well as mass graves.

504th Bombardment Group, USAF, over Mount Fuji, 1945.

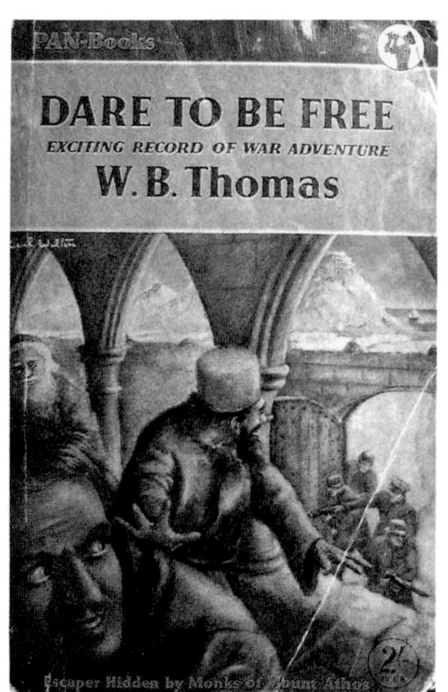

Cover of the first edition of W. B. Thomas's *Dare to Be Free* (1951).

A few metres away from the Grappa sanctuary, a bronze statue pays tribute to the 1,500 partisans executed by the Nazi-Fascist troops on the mountain and in the nearby town of Bassano. In southern Spain a small marble stele on the side of a scarcely trafficked road winding up the Sierra silently commemorates the dozens of republicans massacred during the Spanish Civil War. Their bodies were dumped in a ravine between the villages of Orgiva and Lanjaron and still lie there, covered as they are by the scrub and scented wild herbs. Likewise, the mountains and ravines of northern Greece still resound with the atrocities of the Nazi occupation and of the bloody civil war that shattered the country between 1946 and 1949.

But in some instances mountains have also represented gateways to freedom – and to life. Mount Athos, for example, became a refuge and privileged escape route for Allied soldiers fleeing Nazi prisoner-of-war camps. Except for occasional controls, the Germans did not interfere much with the Athonite monks, who generally empathized and secretly collaborated with the emaciated and often wounded refugees. Disguised as local peasants or even as monks, the refugees took advantage of the craggy topography of the mountain peninsula and its thick forests to elude the Nazis and rejoin their troops in neutral Turkey and Allied Egypt. After the war some of their extraordinary accounts became populars. W. B. Thomas's *Dare to Be Free* (1951), for example, describes the adventurous escape of the New Zealand officer from the Thessaloniki prison camp and the long period he spent hiding on Athos, before crossing the Aegean and rejoining his troops. In his *Escape to Live* (1947), the British Wing Commander Edward Howell describes Athos, the Holy Mountain of Orthodoxy, as 'symbolic':

> [We] had to pass through it and out into the unknown . . . Some would linger in the gateway on the brink of discovery, hesitating to launch out into the unknown, fearful of the

> steps to freedom though eager to attain it. They would see in the lonely places [of Athos] the pale reflection of their goal. Others, however, would come to this place in their spirits and, seeing the road ahead, would step out boldly to savour the full adventure of freedom itself. These men would pioneer the trail for millions; they would shape the future.[45]

Mountains, life and death in the *Bergfilm*

Life and death on mountains have inspired a vast quantity of movies, ranging from films about mountaineering, adventure and exploration to war tragedies and journeys of self-discovery.

Arnold Fanck, *Storm Over Mont Blanc* (1930).

Caspar David Friedrich, *Wanderer above the Sea of Fog*, 1818, oil on canvas.

Nowhere, however, is the interplay between mountain scenery, life and death dramatized with the same poetic intensity as in the Alpine cinematography of the 1920s and 1930s. In the German *Bergfilm*, a genre characterized by 'a combination of auratic landscapes, breathtaking atmospherics and high-pitched emotions', mountains are no mere backdrop, but actual members of the cast. Inaccessible summits, treacherous crevasses, swirling seas of clouds, vertiginous precipices and idyllic high pastures reflect the drama of human passions and of inner conflicts which usually end in tragedy. They become 'dramatic elements, living companions' to climbers, guides and comrades.[46]

The genre, most popular in Weimar Germany, was pioneered and popularized by Arnold Fanck. His mute movies blend high-altitude locations with the melodrama of love affairs and violent deaths. Alpine peaks are invested with an uncanny and monstrous potential; they exude a seductive energy and trap humans into an inescapable dependence. The silent world of high altitude imparts life lessons, tests physical and psychological strength, helps build up virtues and yet at the same time demands the sacrifice of life. Drawing on the aesthetic conventions and iconography of nineteenth-century Romantic landscape painters, such as Caspar David Friedrich, Fanck imbues mountain scenes with a mystical power. His sublime peaks and gorges become projections of incontrollable inner forces.

The Holy Mountain (1926), a film dedicated to a mountain climber friend of Fanck who perished in the First World War, features the story of an expert mountaineer, Robert, his best friend, Vigo, and a charming dancer, Diotima (interpreted by debuting actress and expert climber Leni Riefenstahl, who later became one of the most famous German film directors and Hitler's premier hagiographer). Diotima's performance at the Alpine Grand Hotel entrances the two men. Seeking to 'gain control over the overwhelming impression' Diotima produces on him, after the show Robert rushes to climb a wild summit and retreats into its silence. The two subsequently end up having an affair and their romance intertwines with idyllic high pastures and soft, white snowfields. The mountaintop, nevertheless, remains the man's domain. 'It must be beautiful up there', says Diotima to Robert. 'Beautiful, hard, and dangerous', he responds. 'And what does one look for up there, in nature?' 'One's self.'

On his return from another excursion, Robert witnesses a traumatic spectacle: Diotima embracing another man. Out of shock, he decides to climb the stormy north face of Mount Santo and draws Vigo with him on his climactic pilgrimage. As the two men progress towards the top, 'the walls come alive with the flow of the avalanches'. A night blizzard eventually traps them on the wall. On this occasion Robert learns that

Original poster of Fanck's *The Holy Mountain* (1926).

DER HEILIGE BERG
Regie und Manuskript:
Dr. Arnold Fanck
HAUPTDARSTELLER: LENI RIEFENSTAHL /
LUIS TRENKER / FRIDA RICHARD
BAUTEN: LEOPOLD BLONDER
UFA-FILM IM VERLEIH DER PARUFAMET
UFA
PARUFAMET

Original poster for Leni Riefenstahl's *The Blue Light* (1932).

the man he saw with his beloved was Vigo, who by accident steps into the abyss. Unable to secure him on the icy wall and lift him up, Robert holds the cord on which his friend's body dangles the whole night, refusing to let it go and at least save his own life. Strained and frozen, Robert eventually has to surrender to the mountain. In a desperate act, he decides to sacrifice himself together with his friend. The two disappear into the abyss. 'Above all looms a Holy Mountain', the ending screen title proclaims, 'a symbol of the greatest values that humanity can embrace: fidelity – truth – loyalty – faith'.

The sacrifice of human life in the name of comradeship is also the key motive of *The White Hell of Pitz Palu* (1929). Located between Italy and Switzerland, the peak is haunted by the solitary presence of Dr Johannes Krafft, who since the loss of his wife in an avalanche continues to silently wander on the mountain. Four years after the accident he meets a young couple, Hans and Maria (again, Leni Riefenstahl), who offer to accompany him on his next climb. The tragedy is nearly repeated, as Hans is hit by an avalanche and risks death. The three remain trapped on the mountain, but the couple manages to survive the cold thanks to Johannes, who takes off his jacket and wraps it around Hans to prevent him from freezing to death. Johannes crawls away to an isolated ice ledge and dies.

In *The Blue Light* (1932), the first film Leni Riefenstahl directed herself, tempestuous snowy peaks are substituted with idyllic scenes. The movie shifts away from the western Alps, the theatre of the 'golden age of mountaineering', to embrace the jagged Dolomites. These mountains were only 'discovered' by the British (and subsequently by Mitteleuropean tourists) in the nineteenth century and came to embody a 'gentler', more feminized version of their western counterparts. Until the First World War the Dolomite peaks were usually linked to a quest for picturesque sceneries and Renaissance artistic heritage (for example, the home of Titian in Cadore), rather than to the search for sublime emotions and masculine heroic feats.[47] In Riefenstahl's film the mountain summit nevertheless retains a mystic aura and its deathly agency, as dozens of young men from a (fictional) nineteenth-century village at the foot of Monte Cristallo (near Cortina d'Ampezzo) fall from its treacherous slopes, one after another. Their fatal climbs are triggered by a glowing blue light that exudes from a crack in the mountain, a quirk of nature caused by crystals illuminated by a full moon, and seems to exercise a hypnotic attraction on them.

Blame is cast on the female protagonist, Junta, who is believed to be a witch. Mistreated by the superstitious villagers, the girl lives in the solitude of the surrounding mountains. On full moon nights, she climbs Cristallo to reach her sanctuary, a grotto covered with beautiful crystals, from which the blue light emanates. Yet life ends in tragedy for her as well, when a painter from Vienna who has fallen in love with her follows the girl to the grotto and, having realized the immense financial potential of the crystals, instructs the villagers how to reach it. When Junta finds her sanctuary barren of the precious stones, she falls to her death. Her mystical figure becomes 'an image and a commodity, a kitsch object hawked by children to tourists, a face framed by crystals which also adorns the cover of the written version of the popular village tale'.[48]

As he is awaiting death in his bus at the feet of the Alaskan mountains, Christopher McCandless tears the final page of Louis L'Amour's biography. It features the following lines from a poem, 'Wise Men in their Bad Hours', by Robinson Jeffers:

> Death's a fierce meadowlark: but to die having made
> Something more equal to centuries
> Than muscle and bone, is mostly to shed weakness.
> The mountains are dead stone, the people
> Admire or hate their stature, their insolent quietness,
> The mountains are not softened or troubled
> And a few dead men's thoughts have the same temper.

Silent, immobile and unforgiving, mountains take human lives. Yet at the same time they also allow humans to put life into perspective. This, however, presupposes a specific way of perceiving space, time and nature, which the next chapters will explore.

4 Mountains and Vision

Climb mountains to see lowlands.
Chinese proverb

Mountain encounters have helped define particular ways of seeing, experiencing and representing the world. The lives and deaths of legendary mountaineers and hermits have been moulded through the intimate contact with the rock of the faces they climbed or the caves in which they dwelled. Holy mountains have also traditionally bounded the visual and imaginative horizons of different civilizations, and formed the *axes mundi* and damned peripheries of others. Throughout human history peaks have functioned as reassuring landmarks for orientation – from ancient Greeks hopping from island to island in the fragmented Aegean to Viking seafarers sailing down the jagged coasts of Norway. Mountains have also functioned as signalling stations, and examples are found in ancient Greece and Mesopotamia, as well as in the Roman and Byzantine empires and medieval Scandinavia. Not least, mountains have served as landmarks for bombing, as in the case of Fuji during the Second World War.

Mountains are not only the most visible landmarks in the landscape, they are privileged vantage points. The ancient Greeks and Byzantines referred to mountain heights as *skopiai* and Romans as *speculae*, meaning 'lookout places'. The Greek poet Simonides (556–468 BC), for instance, speaks of the summits of Cithaeron as 'lonely watchtowers', whereas Strabo (64 BC–AD 24) provides descriptions of an actual belvedere built on one of the summits of Mount Tmolus in Lydia (western Anatolia).[1]

Lucian of Samosata (AD 125–180) dedicated an entire satirical poem to the view from the mountaintop. In his 'Sightseers',

Charon, who has been given a day's holiday from his usual job as ferryman of the dead, asks Hermes 'to explain to him the unfamiliar sights of the world of the living'. Hermes agrees and decides that a high peak would make for a suitable spot from which to survey the earth. The two thus pile Ossa and Pelion on Olympus, and then add Oeta and Parnassus on top of them. From the height of the mountain pile they survey human lives and fortunes, but only to find out 'the small space Charon – that is, Death – seems to occupy in the thoughts of those who play their parts on this stage'.[2]

With the exception of gods and semi-divinities, the most common mountain climbers of antiquity were nevertheless rulers and wise men. Even before Christ was taken by the Devil to the top of the Mount of Temptation, Philip the Macedon ascended the highest peak of the Haemus range in order to see the lie of the land as he planned his war against Rome. (It was widely believed that this summit commanded a view over the Danube and the Alps and both the Adriatic and the Black seas.)[3] Likewise, Hadrian ascended Etna in Sicily and Mount Casius in Syria to observe the sunrise and obtain a view over a wide swathe of country, whereas Atlas, the legendary ruler of Mauretania, who was also a philosopher, astronomer and mathematician, was said to have climbed up to the highest summit in his kingdom to gain a prospect of the entire world – like Charon and Hermes.[4]

From antiquity to the present day, gazing from a mountaintop has been traditionally interpreted as an empowering act – as a supreme expression of political authority and knowledge. The colonial gaze of European explorers, like Mackinder, represents, perhaps, the most emblematic example in this respect. Ultimately, however, the view from the mountaintop embeds the fundamental tension discussed in the previous chapters. As a metaphor of omniscience, it is at once divine and diabolic; it allows control over the landscape and at the same time it causes dangerous vertigo, as it did for the Austrians on Mount Grappa. Modernity is defined precisely through the tension between these two poles. Yet can we really talk about *a* modern way of seeing and representing the world? Did premodern cultures and do non-Western cultures see mountains and see *from* mountains in the same way we do? How have specific mountain encounters shaped specific ways of seeing the world?

Topographies of memory: Mount Nebo

In about AD 384 a pious Spanish woman called Egeria, probably a nun, stood on the top of Mount Nebo, the 'lofty peak' Moses ascended at the end of his life to behold the Promised Land (Deut. 32:49–50). Having performed the customary prayers and read the relevant passage of the Bible in a small church on the summit, Egeria was invited by the local clergymen to go out and behold 'the places which are described in the Books of Moses'. From the height of her panoramic platform the pilgrim saw a giant topographic map of biblical places unfold before her eyes: the Jordan running into the Dead Sea and Jericho on the far side, jutting out over the valley; the city of Zoar and the country of the Sodomites; the Promised Land and everything in the area of Jordan 'as far as the eye could see'.[5]

Egeria had travelled from the other edge of the empire (and of the known world) to behold the sites of the Bible, as she said, 'with her own eyes'. Her journey through the Holy Land was signposted by high places – craggy hills, high cliffs, dramatic gorges opening on wide prospects. The pilgrim was continuously seeking elevated vantage points, or *specula*, from which she could gain commanding views. Mountains like Nebo served Egeria both as *loci memoriae*, that is, as landmarks activating biblical memory (for example, the death of Moses), and as panoramic platforms from which to navigate the landscape underneath.

On the top of Mount Nebo, the clergymen guided Egeria's eye through biblical places. They pointed at nearby and distant cities; at more and less prominent landmarks, ranging from the spot where the wife of Lot was turned into a pillar of salt to dramatic peaks, such as Agri Specula, or Viewpoint.[6] Thus directed, Egeria's gaze pushed her mind where her body could not reach. Unlike traditional pilgrims, content with the

identification of a stone or a tomb as they were, Egeria extended her field of vision far beyond specific sites. Her eye oscillated between the topographic detail and the horizon, between what was there and what was yet to come. From its elevated position, it hopped from one hill to the next; it ran through long valleys, followed the course of rivers, explored modern and ancient settlements, scrutinized distances.

Mount Nebo set Egeria in a position of mastery over territory. Her view was the view Christ rejected on the Mount of Temptation, but the one ancient philosophers and rulers longed for – from the imaginary extraterrestrial flights of Socrates and Cicero to King Atlas' global gaze from the top of his Mauretanian mountain; from Philip's ascent of Haemus to Alexander the Great's ascent to heaven to behold his conquests.[7] Unlike the wise men and monarchs of the past, Egeria, however, privileged totalizing bird's-eye views, because they granted her mastery not over territory but over biblical topographies and ultimately over the biblical past. Each location evoked a story, a scriptural passage.

View from the retaining wall of the Byzantine monastery on Mount Nebo.

From the height of Tabor, Sinai, Nebo and other famous peaks, Egeria was presented with vast memory theatres in which biblical narratives were activated through her mobile eye. Such views are accurately described in a letter she sent to her sisters who had remained home in Spain. As she explained, the mental visualization of those places would enable them to memorize better the Bible.

Allegorical visions: Mount Ventoux

If faith and topographies of memory underpin Egeria's mountain hopping, curiosity and aesthetic gratification motivate the most famous mountain climb in Western intellectual history, that is, Petrarch's ascent of Mont Ventoux, the highest peak in Provence, at 1,912 m. Almost a millennium after Egeria, the Italian poet decided to climb the mountain, which had been haunting his imagination for many years. 'I have lived in this region from infancy', he wrote to Dionisio da Bergo San Sepulchro, an Augustinian monk at the University of Paris,

> Consequently the mountain, which is visible from a great distance, was ever before my eyes, and I conceived the plan of some time doing what I have at last accomplished to-day. The idea took hold upon me with especial force when, in re-reading Livy's History of Rome, yesterday, I happened upon the place where Philip of Macedon, the same who waged war against the Romans, ascended Mount Haemus in Thessaly, from whose summit he was able, it is said, to see two seas, the Adriatic and the Euxine.[8]

As Petrarch confesses to the monk, his only motivation was 'the wish to see what so great an elevation had to offer'. The climb nevertheless soon turns into an allegory of his life journey, as the youngster repeatedly tries to find an easier way to the top, but each time his route turns out to be lengthier and more straining than the direct path chosen by his brother. 'I was simply trying to avoid the exertion of the ascent; but no human ingenuity can alter the nature of things, or cause anything to reach a height by going down.' Disgusted by the intricacy of his detours, Petrarch summons himself that 'thou must perforce either climb the steeper path [of life], under the burden of tasks foolishly deferred, to its blessed culmination, or lie down in the valley of thy sins, and (I shudder to think of it!), if the shadow of death overtake thee, spend an eternal night amid constant torments.'[9]

Stimulated by these thoughts, Petrarch rejoins his brother and eventually reaches the highest point of the mountain. The 'great sweep of view' unfolding before him combined with the 'unaccustomed quality of the air' initially causes a sense of vertigo. 'I stood like one dazed. I beheld the clouds under our feet, and what I had read of Athos and Olympus seemed less incredible as I myself witnessed the same things from a mountain of less fame.'[10]

However, as his gaze turns southeast, to Italy, the poet once again moves from external to inner contemplation and recalls the years he has left behind and his boyhood studies.

The mountaintop offers him a bird's-eye view not only of the surrounding area, but of his life itself:

> The Alps, rugged and snow-capped, seemed to rise close by, although they were really at a great distance; . . . I sighed, I must confess, for the skies of Italy, which I beheld rather with my mind than with my eyes. An inexpressible longing came over me to see once more my friend and my country . . . Thus I turned over the last ten years in my mind, and then, fixing my anxious gaze on the future . . . I rejoiced in my progress, mourned my

View from Mont Ventoux.

weaknesses and commiserated the universal instability of human conduct.[11]

View of Mont Ventoux from the south.

As if awakened from sleep, Petrarch suddenly recalls the initial purpose of his ascent. Having cast homesickness and lovesickness aside, he thus turns his gaze to the west and launches into a detailed exploration of the panorama. As passions and anxieties dissipate and his organism adjusts to the high altitude, the landscape acquires crisper contours. He is able to identify the bay of Marseilles and the mountains of the region about Lyons (though these remains generic mountains, rather than named peaks).

His reconnaissance complete, Petrarch pulls from his pocket a copy of St Augustine's *Confessions* and decides to read at random. His eye falls on the following passage: 'And men go about to wonder at the heights of the mountains, and the mighty waves of the sea, and the wide sweep of rivers, and the circuit of the ocean, and the revolution of the stars, but themselves they consider not.' At this point, bodily vision is once again superseded by spiritual insight. Ashamed, the young man turns back to the valley. 'I had seen enough of the mountain. I turned my inward eye upon myself.'[12] Ultimately the mountaintop enabled Petrarch to put his passions and terrestrial life 'into perspective'.

Petrarch's ascent has been taken as emblematic of a novel sensibility. His ascent is an allegorical as much as physical feat. While Egeria climbed Nebo and other biblical peaks after Moses and used sight to validate and better memorize biblical truths, Petrarch is caught between Philip the Macedon's thirst for terrestrial omniscience and St Augustine's acknowledgement of the limit, or rather deceitfulness, of terrestrial things. 'Could the survey of the outer world (and what better place to seize its form than from the prospect of a mountaintop?) ever disclose essential inner truth?'[13]

Ultimately Petrarch's venture remains a typically Christian inner struggle followed by epiphany and redemption (the reading of St Augustine's passage). On the mountaintop the poet overcomes homesickness, lovesickness and melancholy in the

Duccio da Buoninsegna, *The Temptation of Christ on the Mountain*, 1308–11, tempera on poplar panel.

same way as Anthony the Great defeated the demons and Christ himself resisted Satan's allure. The Mount of Temptation cast its long shadow on Mont Ventoux – and on the history of Western vision.

Putting the world into perspective: back to the Mount of Temptation

Shortly before Petrarch's ascent, Duccio di Buoninsegna, an artist from Siena, painted one of the few extant medieval representations of Christ's *Temptation on the Mount*. Here the lofty peak nearly disappears under Christ's mighty presence. Eternity, suggested by the gold leaf on the higher part of the composition, stands in contrast with the fabulous yet ephemeral walled cities in the lower part. Christ is firmly standing on the mountain while Satan is stumbling on his illusory cities. The mountain features as a boulder – a visual reminder of Christ's solid rock of faith, as opposed to the futility of mundane affairs.

The painting is ruled by memory and symbols. The sizes, locations and colours of the figures are proportionate to their significance in the Bible and the need to commit them to memory. They do not respond to the geometrical principles of linear perspective, but to the power of memory. Hence the viewer's attention is immediately captured by Christ at the centre of the composition and subsequently moves to other individual features: the Devil, the angels, the cities, the boulder-like mount. This technique was common in Byzantine painting but it also (more broadly) reflects a typically premodern topographic way of seeing.

The same scene, 500 years later, assumes very different contours. In an engraving by William Richard Smith (1829), scenery takes over allegory. Here Christ and the Devil nearly disappear in the landscape. What is alluring is not the beauty of the cities but the overpowering cartographic view from above – and the infinite horizon. As opposed to Duccio's painting, in

William Richard Smith, *The Temptation on the Mount*, 1829, engraving.

Smith's engraving landscape is no longer an ensemble of places (or *loci memoriae*), but a scenery mastered by a monofocal gaze from a fixed point of view. The eye is guided through a vertiginous rocky platform to Christ and the Devil, at the convergence of visual axes. At the same time the eye is also pushed to the horizon. Landscape is articulated through such tension – between proximity and infinity.

According to the Italian geographer Franco Farinelli, the modern idea of landscape was born only when mountains were measured and the maximum horizon defined; that is, not long before this engraving was printed.[14] This shift in the perception and representation of landscape, however, has deeper roots. Before aerostatic balloons were sent into the air, hills and summits constituted the privileged strategic viewpoints for planning battles and towns alike.

Comprehensive bird's-eye views, as if witnessed from a mountain or a hilltop, are a common feature in Renaissance painting. Leonardo da Vinci has been especially credited for advancing aerial imagination, and landscapes seen from above, through a high oblique angle, form the background of several of his paintings. Such views, perhaps inspired and facilitated by the hilly topography of central Italy, are consistently signposted by mountains, a testimony to Leonardo's fascination with these objects of nature.[15] From his early works to the *Last Supper*, the *Mona Lisa*, the *Madonna Litta* and the *Virgin of the Rocks*, hazy peaks loom on distant horizons through windows, porches or natural openings. The representation of mountains, Leonardo argued, deserved special care:

> O Painter! When you represent mountains, see that from hill to hill the bases are paler than the summits, and in proportion as they recede beyond each other make the bases paler than the summits; while the higher they are, the more you must show of their true form and colour.[16]

Mountains are at once settings and objects for artistic inspiration. Yet Leonardo's fascination with mountains is scientific, rather

Leonardo da Vinci, *Virgin of the Rocks*, 1485, oil on poplar.

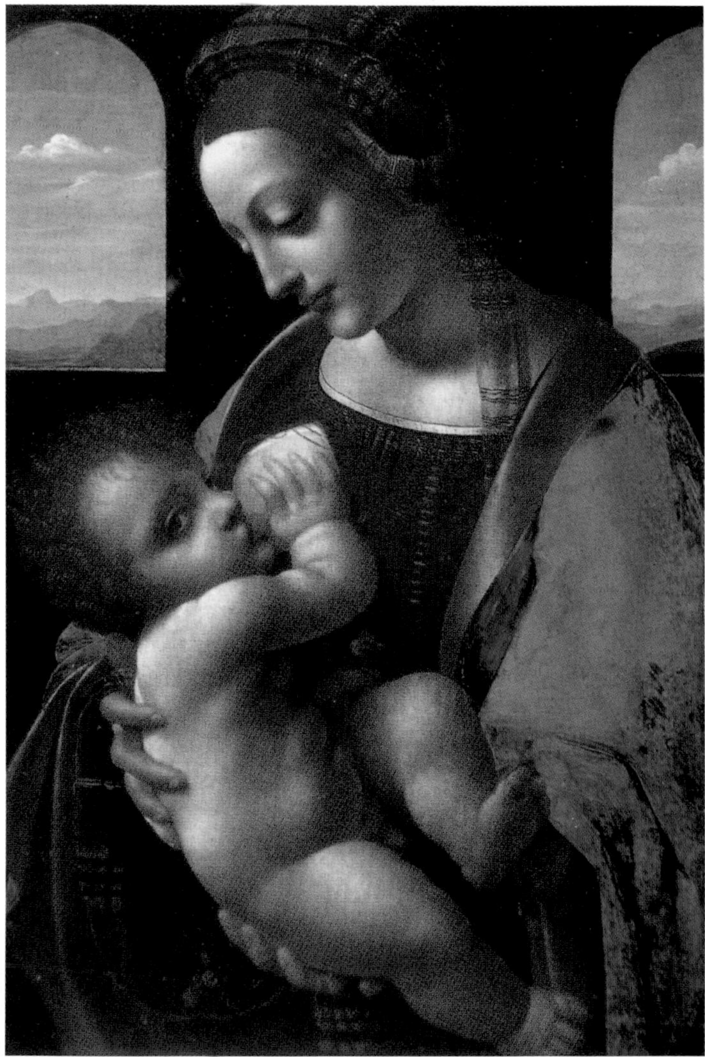

Leonardo da Vinci, *Madonna Litta*, 1490, oil on canvas.

than solely aesthetic. The artist speculates on their formation, considers their composition and climbs to their top to study atmospheric thickness. His notebooks alone contain the word 'mountain' over 200 times.

In the sixteenth century aerial views became popular not only as backgrounds (as in Leonardo's works) but as subjects in themselves. The so-called 'cosmographic paintings' set the observer in an elevated position (usually a highest mountaintop or cliff) and offered a vast panorama of impossibly distant places with multiple viewpoints – as with Christ on the Mount of Temptation.

Pieter Bruegel's *Fall of Icarus* (*c.* 1558), for example, brings both Crete and Cyprus within visual reach. In Albrecht Altdorfer's *Battle of Alexander at Issus* (1529) landscape assumes an even vaster extent. The whole Eastern Mediterranean becomes the stage for the global drama of the defeat of the Persian army in 334 BC. Dazzling numbers of armed soldiers pour out of walled cities and encampments framed by distant lands and seas: the Levantine coast, Cyprus, the Isthmus of Suez, the Nile Delta and the Red Sea stretching to the curved horizon. The eye is raised to a position where 'the site of battle, the curving earth and the planetary bodies are all brought within its scope.'[17]

As landmarks, mountains add further drama to cosmographic paintings. In Bruegel's *Fall of Icarus* distant summits frame the sun rising on the sea and reinforce the sense of terrestrial curvature, whereas in Altdorfer's *Battle of Issus* the magnitude of the stylized peak brutally looming over the ocean of soldiers seems

Pieter Bruegel the Elder, *Fall of Icarus*, *c.* 1558, oil on canvas.

to extend into the sky, as the contours of the rocks are echoed in the shapes of the cosmic swirl of clouds.[18] As vantage points, mountains operate as liminal spaces setting the viewer between elemental and celestial spheres. They offer to every observer 'the dignity and authority over space once reserved for gods and monarchs'; at the same time they also cause exhilaration and dizziness.[19]

The apotheosis of such cosmographic vision is marked by Milton's literary representation of Christ's temptation on the mount, probably the most famous and evocative of all renditions of the biblical passage. The reader of *Paradise Regained* is taken with Jesus up to the top of the highest mountain and offered a dazzling view of the plain below and of the entire cosmos akin to Altdorfer's epic panorama:

> Huge cities and high-towered, that well might seem
> The seats of mightiest monarchs; and so large
> The prospect was that here and there was room
> For barren desert, fountainless and dry.[20]

Past and present glories converge in a single moment and landscapes that cannot be seen by the mortal eye are brought together in a single bird's-eye view. From the mountaintop the reader beholds with Christ

> Assyria, and her empire's ancient bounds,
> Araxes and the Caspian lake; thence on
> As far as Indus east, Euphrates west,
> And oft beyond; to south the Persian bay,
> And, inaccessible, the Arabian drouth.[21]

A succession of famous cities parades under Christ's and the reader's gaze: Nineveh, Babylon, Persepolis, Bactra, Hecatompylos, Susa, Seleucia, Nisibis, Artaxata, Teredon, Ctesiphon – and the list continues. Armies and soldiers come next, outpouring from city gates in the same 'numberless numbers' as in Altdorfer's painting:

Albrecht Altdorfer, *The Battle of Alexander at Issus*, 1529, oil on wood.

ALEXANDER M DARIVM VLT SVPERAT
CAESIS IN ACIE PERSAR PEDIT C M EQVIT
VERO X M INTERFECTIS MATRE QVOQVE
CONIVGE LIBERIS DARII REGIS CVM M HAVD
AMPLIVS EQVITIB FVGA DILAPSI CAPTIS

A multitude, with spades and axes armed,
To lay hills plain, fell woods, or valleys fill,
Or where plain was raise hill, or overlay
With bridges rivers proud, as with a yoke:
Mules after these, camels and dromedaries,
And waggons fraught with utensils of war.[22]

From the Mount of Temptation to Mont Blanc: modernity and the panoramic view

With the notable exception of Petrarch, Leonardo and a few others, medieval and Renaissance views from above remained mostly epic flights of the human imagination. In the eighteenth century, as the peaks of Western Europe were starting to be systematically climbed and measured, the view from the mountaintop underwent a further transformation: it became a view constructed around and *for* a sovereign subject standing alone and 'first' on the summit.[23]

In 1787 Horace-Bénédict de Saussure, a professor of natural philosophy from Geneva, pompously claimed to have 'conquered' the summit of Mont Blanc, the highest peak in the Alps. Two locals, Jacques Balmat and Michel Paccard, had accomplished the first successful ascent of the mountain in the previous year, claiming a reward Saussure himself had promised for the enterprise. Saussure has nevertheless been commonly credited as the true conqueror of the mountain, and even as the man responsible for a new aesthetic perception of the Alps. Why was this the case?

Unlike Balmat, a humble peasant labourer and crystal hunter, and Paccard, a stout physician with scarce knowledge of mineralogy, Saussure was a professional scientist. He had crossed the Alps no fewer than fourteen times and extensively studied their geology. To Mont Blanc he took a considerable amount of scientific equipment and personal belongings, including a tent and a portable stove – all carried by a retinue of nineteen porters. The goal of his sensationalized climb was not merely to reach the highest point of the mountain, but 'to make the observations and experiments which alone would give value to [the feat]'; in other words, to transform the peak from cursed wasteland into an object of science.[24]

More specifically, Saussure believed that Mont Blanc could cast light on the formation of the earth. He wrote that the peak 'seems to be the key of a great system, but unfortunately it is scarcely accessible'.[25] High Alpine peaks allowed the naturalist to 'embrace at once a multitude of objects . . . The eyes, at once dazzled and drawn in every direction, know not at first where to fix themselves. Little by little the eye adjusts and selects its objects of study', thus uncovering hidden orders.[26]

While aesthetics was not a key concern for Saussure, his accounts dramatically contributed to shaping a new way of seeing. Ultimately it was the naturalist's ordering gaze, his omniscient view from the mountaintop, that enabled the appreciation of Western Europe's high places. Unlike Egeria's eye wandering through biblical memory places, or Petrarch's, satisfied as he was with a basic reconnaissance of generic mountains, Saussure's eye was after the order in nature. His was a holistic, totalizing view. On the top of Mont Blanc he could enjoy the grand spectacle which lay beneath his eyes:

> A light vapour suspended in the lower regions of the air robbed me of the sight of the lower and more distant objects, such as the plains of France and Lombardy; but I did not much mind this loss. What I saw and saw with the greatest clearness, was the whole collection, the whole group of these high peaks of which I had so long desired to know the organization. I could not believe my eyes; it seemed to me that it must be a dream when I beheld beneath my feet those majestic peaks, those veritable needles, Le Midi, l'Argentière and the Géant, whose bases even I had so long found difficult and dangerous of access. I seized on their bearing one to another, their connection, their structure; and one glance removed all those doubts which years of labour had not been able to clear up.[27]

From the valley Saussure's wife and sister anxiously followed the course of his ascent through a telescope:

> [As I reached the peak] my eyes were first turned towards Chamounix . . . I experienced a very sweet and consoling feeling when I saw floating in the air the flag which they had promised to hoist the moment when they espied me on the highest point, and when their fears would be at least relieved for the time.[28]

For those who followed it from a distance, the ascent was turned into a spectacle consumed through a mechanized gaze; for the scientist on the mountaintop, the world itself had become a vast spectacle constructed around him.

This way of seeing is best represented by the fisheye view. Eight years before the legendary ascent, during his geological explorations, Saussure sketched the panorama he had enjoyed

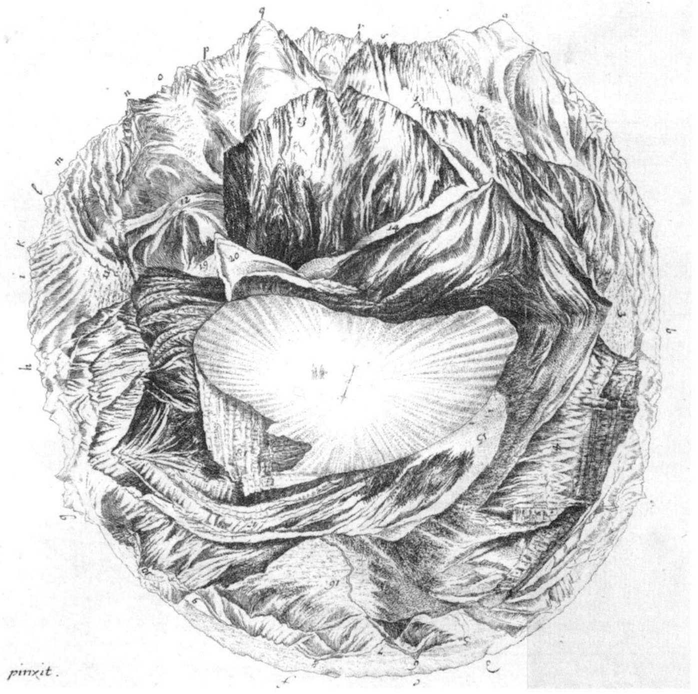

Horace-Bénédict de Saussure and Marc-Théodore Bourrit, *Circular View of the Mountains from the Summit of the Buet*, 1779.

on the summit of Mount Buet (3,096 m) and commissioned Bourrit to produce one such view of the surrounding mountains. As he explained, 'the spectator is placed at the centre of the image and all the objects are drawn in perspective from this centre, as they would present themselves to an eye situated at the same centre which successfully made a tour of the horizon'. The image conveys the sense of unity and panoptic control Saussure was after (and eventually achieved on the top of Mont Blanc), as well his own centrality. 'The illustrator draws the objects exactly as he sees them by turning his paper as he turns himself . . . Thus [one sees] successively all the objects linked together precisely as they would appear to an observer situated on the summit of a mountain.'[29]

In the nineteenth century fisheye views gained momentum. Their applications transcended science; they became part of a new popular visual culture. For example, they were used as maps for orientation in the panoramic rotundas of the great European

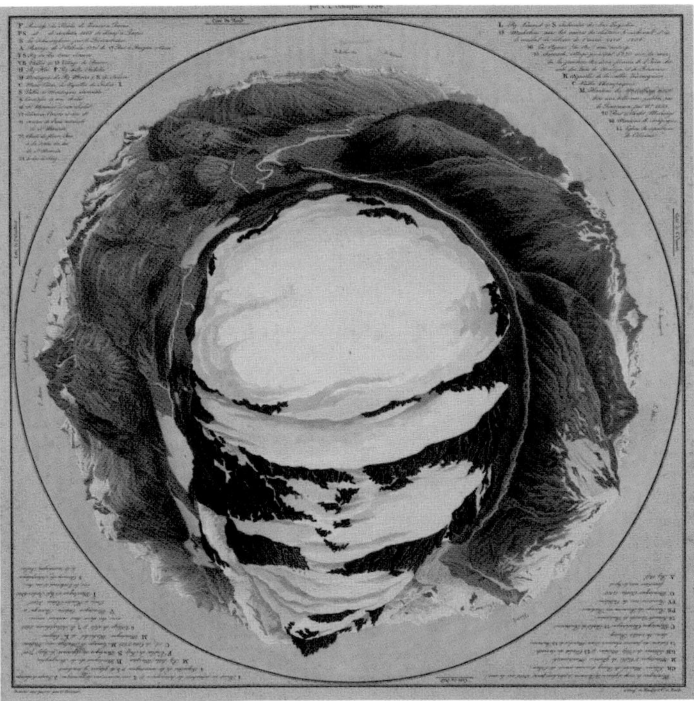

Elias Emanuel Schaffner, 'Panorama des Alpes Rhétiennes du Haut-Engadin', 1836.

capitals. These were windowless circular structures in which the viewer could admire a vast 360-degree panoramic painting from a platform set at the centre of the building. The painted canvas entirely covered the walls and displayed urban and natural landscapes, including mountain landscapes, as well as battle scenes and other heroic feats. The visitor's field of vision was cut by a coverage above the platform and by the platform itself. This produced the illusion of total immersion in an actual landscape – and in fact panoramas were often sold as surrogates for travel. Visitors could freely wander on the platform and take advantage of specialized 'guides' who would point at the various features in the landscape and identify such features on fisheye maps.

The first panorama was invented by an Irish painter in 1785 (two years before Saussure's conquest of Mont Blanc) and exhibited in London eight years thereafter. In less than a decade, a true 'panoramania' exploded throughout Europe and beyond. The Prussian naturalist and explorer Alexander von Humboldt encouraged the construction of public rotundas, 'containing alternating pictures of landscapes of different geographical latitudes and from different zones of elevation'. These spaces were meant to familiarize the public with 'the works of creation' and their 'exalted grandeur'.[30] While Romantic poets and painters insisted on the impossibility of total knowledge, and adopted mountains as 'a surrogate for the indefinable and indescribable elements of the human psyche', panoramas, alongside three-dimensional mountain models, promised complete knowledge of a landscape, an 'all-seeing eye'.[31]

As with Saussure's mountaintop views, panoramas portrayed the world as an ordered totality centred on the individual. Yet, while producing a liberating optical effect (a world constructed around the viewer for their own visual consumption), panoramas also caused vertigo. 'Ladies of a nervous disposition' were repeatedly warned 'to be on their guard'.[32] Such tension – between orientation and disorientation, rational exploratory gaze and suffering body – was repeatedly experienced by Saussure himself during his Alpine wanderings. Climbing to the top of a mountain required the ability to face precipices – in other words, to

Robert Fulton, *Description explicative du panorama ou tableau circulaire et sans borne ou manière de dessiner, peindre et exhiber un tableau circulaire* (Panorama with Observation Platform), April 1799, pencil drawing.

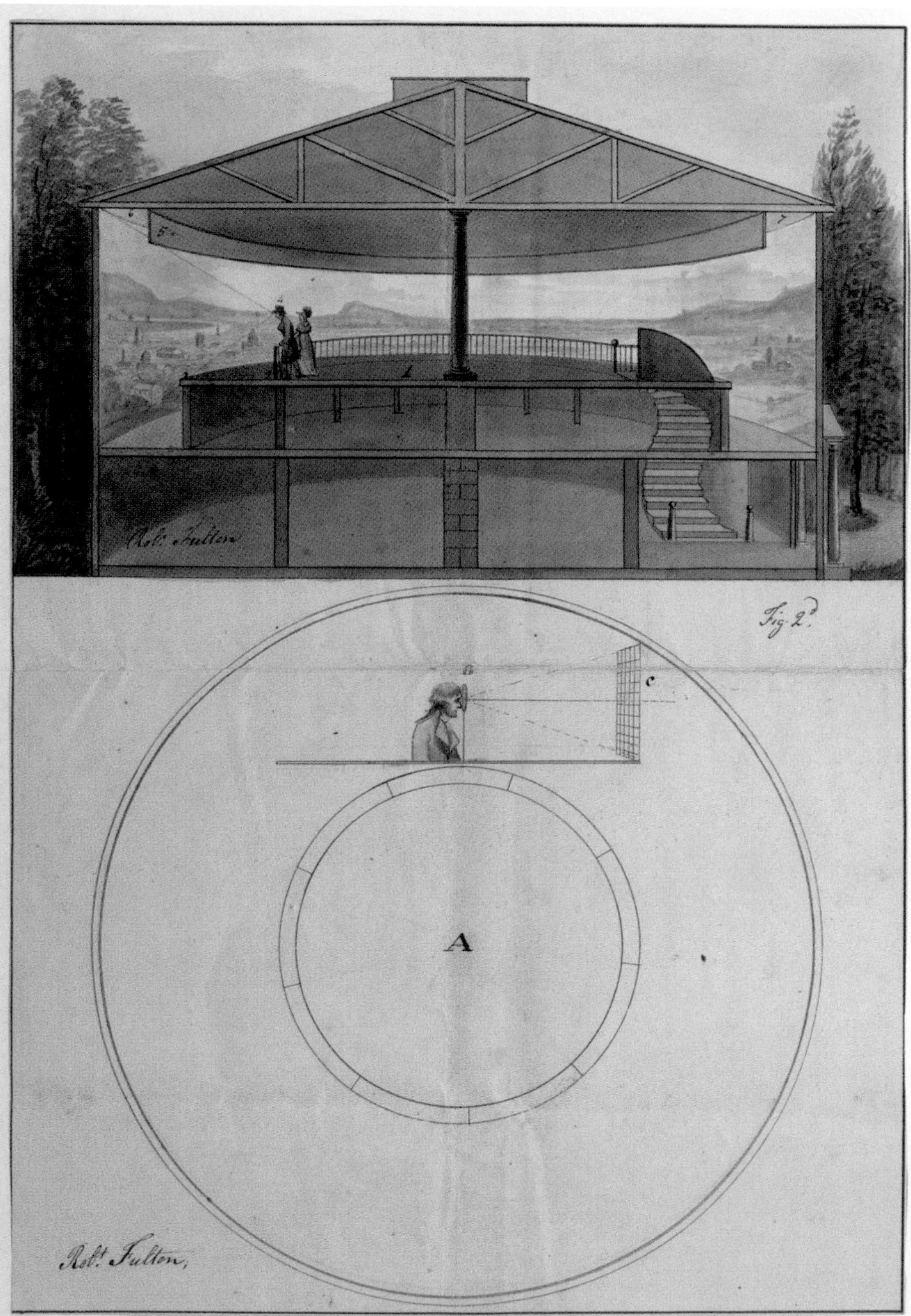
Robt. Fulton
Fig. 2d.
B
C
A
Robt. Fulton,

tame one's senses. Saussure developed a technique: 'lying down and pushing one's head forward until it reaches the edge of the ravine is how to get used to seeing the deepest abysses without fears and vertigo.'[33]

Ironically, the Alps became familiar to the public mainly thanks to the new visual technologies. Lay citizens could comfortably explore the inaccessible high places of Europe in urban rotundas with the expert assistance of professional guides and war veterans, and at dedicated events, such as the Munich Sport Exhibition of 1899, featuring three-dimensional mountain models and panoramas. They could also enjoy Alpine fisheye

War veteran at the panorama of the Storm of St Privat, Berlin, 1883.

Porcelain dishes featuring fisheye views of Mont Blanc, Vienna.

views on fine porcelain dishes and other visual mementoes.[34] From 1852 to 1859 Londoners could also experience the ascent of Mont Blanc through Albert Smith's popular show in Piccadilly. The British satirist had himself climbed the peak in 1851 with a retinue of guides comparable to Saussure's expedition (only instead of scientific equipment they carried extraordinary provisions of alcohol: 66 bottles of *vin ordinaire*, six bottles of Bordeaux, ten bottles of St George, fifteen of St Jean, ten of Cognac and two of Champagne).

The show, an 'extravaganza of Alpine kitsch', featured dioramas of the mountain rolling across the back of the stage and commented on by Smith. Through the combination of his theatrical narrative and visual technologies, Smith aimed at recreating the sense of vertigo and disorientation he had experienced on the mountain:

> For a space of two hours, I was in such a strange state of mingled unconsciousness and acute observation – of combined sleeping and waking – that the old-fashioned word 'bewitched' is the only one that I can apply to the complete confusion and upsetting of sense in which I found myself plunged . . . [On the mountaintop] I wanted the whole panorama condensed into one point; for, gazing at Geneva and

Mr Albert Smith's Ascent of Mont Blanc at the Egyptian Hall, Piccadilly. From the *Illustrated London News*, 25 December 1852.

> the Jura, I thought of the plains of Lombardy behind me; and turning round towards them, my eye immediately wandered away to the Orberland, with its hundred peaks glittering in the bright morning sun. There was too much to see, and yet not enough: I mean, the view was so vast that, whilst every point and valley was a matter of interest, and eagerly scanned, the elevation was so great that all detail was lost.[35]

Smith's audience, who by 1858 was in the hundreds of thousands, could also take the experience of displacement home, thanks to portable visual technologies such as peepshows.

Sometimes criticized by professional mountaineers as vulgar, Smith's show nevertheless generated a true 'Mont Blanc mania' in Victorian Britain. Between 1853 and 1858 the mountain was ascended no fewer than 88 times. The editors of *Punch* sarcastically reported that the route to the top was to be carpeted. By the end of the century 'certificates announcing that one had reached the summit of Mont Blanc were almost as easy to find as the sight of people trudging up and down its sometimes-dangerous routes.' At the same time, observing climbers through telescopes had become a sport as popular as mountaineering.[36]

Not only had visual technologies turned mountains into commodities for the consumption of the European urban bourgeoisie, but the view from the mountaintop transformed the world itself into an exhibition centred on a spectator paradoxically immersed and yet at the same time detached from it – as in the panorama. 'Marvellous some of the panoramas seen from the greatest peaks undoubtedly are; but they are necessarily without those isolated and central points which are so valuable pictorially,' wrote Edward Whymper.

> I think that the grandest and the most satisfactory standpoints for viewing mountain scenery are those which are sufficiently elevated to give a feeling of depth, as well as height, which are lofty enough to exhibit wide and varied views, but not so high as to sink everything to the level of the spectator.[37]

Albert Smith's Mont Blanc peepshow.

Other ways of seeing

There are different reasons for gazing at the lie of the land from a mountaintop: military reconnaissance, pious contemplation, scientific observation, planning, mastering or simply mere curiosity. There are, however, also different ways of looking at the landscape. The perspectival view from the mountaintop has been taken as a metaphor for modernity and its many contradictions. It is the one way of seeing we have come to take for granted. However, we often forget it is just one way of looking at landscape – and at the world.

On Byzantine and Russian icons, for example, mountains are not objects to be mastered through linear perspective, nor are they elevated platforms for mastering the world. Instead, they usually feature as slabs of stone stacked in ascending pinnacles bending towards the centre of the icon. In this way, they concentrate the light at the centre of the composition out towards the beholder standing before it. The gaze of the beholder is reflected back to his or her heart – the vanishing point of the composition. Ultimately, the image masters the beholder, and not the other way around.

Inverse perspective in an icon of Holy Trinity after Rublev, 15th century.

Chinese landscape painting offers a different way of seeing. While Western perspectival painting (and later mechanized photography) requires a fixed viewer gazing down from an elevated vantage point, Chinese landscape painting emphasizes the necessity of moving through the landscape, of wandering through the mountains. In learning to paint, wrote the eleventh-century artist Kuo Hsi, 'you must go in person to the countryside to discover it. The significant aspects of the landscape will then be apparent'.[38] Physical wandering allows the artist to absorb the essence of the landscape. Only when

Kuo Hsi, *Early Spring*, 1072, hanging scroll.

this essence has been assimilated should he attempt to paint it. Unlike in Western painting, distant features are not necessarily smaller, but they are located higher in the composition. 'The artist moves through the very world he will paint. He will paint the vision he has gained while wandering', and in turn will make the viewer wander through the verticalities of the composition.[39]

In Cubist art mountains serve yet another function. In Paul Cézanne's paintings of Montagne Sainte-Victoire the peak

Paul Cézanne, *Mont Sainte-Victoire*, 1902–4, oil on canvas.

towers over the surrounding plain in forceful grey and black brushstrokes, conveying to the image an intrinsic, almost magnetic force. It becomes the pivot of the composition. Cézanne is not a detached viewer but, as he wrote to a friend, 'the landscape thinks itself in me.' Cézanne's paintings of this modest and yet most celebrated peak in Western art have been ascribed to the artist's obsession with the mountain – he produced over sixty paintings of it – as well as to a crisis of perception, 'the recognition that looking at one thing intensely did not lead to a fuller and more inclusive grasp of its presence, of its immediacy. Rather, it led to its perceptual disintegration and loss.'[40] Sainte-Victoire is an anti-Mount of Temptation – the other side of modern vision, or simply another way of seeing mountains and the world.

According to the Scottish novelist and poet Nan Shepherd, writing on the Cairngorms in the 1940s, mountains allow us to see things in new ways. Yet one has to train and discipline the senses, including the eye. As she takes her readers on long walks along dark ridges and crystal lochs in her evocative *The Living Mountain*, Shepherd challenges them to pry through surfaces, to venture into hidden crevasses, to pause over details, to immerse themselves fully in the landscape and all its elements. Her poetic journeys are a perpetual act of discovery. Rather than pursuing the modern dream of omniscience, or passively surrendering to the impossibility of mastering the sublime, she takes the mountain experience as a creative act, an infinite act of learning, an enriching but always unfolding process.

> The more one learns of this intricate interplay of soil, altitude, weather and the living tissues of plants and insect . . . the more the mystery deepens . . . Knowing [the mountain] is endless. The thing to be known grows with the knowing.[41]

In this process, vision plays a prime role.

The view from above entrances Shepherd. From the height of the mountaintop, the world seems 'to fall away all round, as though I have come to its edge and were about to walk over. And far off, on a low horizon, the high mountains'.[42] Unlike Saussure, however, it is not lucidity of image and open horizons that excite her most, but the snowflake's geometrical shapes, the quartz crystal, the patterns of stamen and petal, the crack in the rock, the deep recess, the ascent of the inside of a cloud, the haze hiding and revealing new shapes, a walk in the dark, 'oddly enough, reveal[ing] new knowledge about a familiar place', or the illusions of the eye caused by elevation – in other words, those hidden sides of the landscape.

> The end of a climb meant for me always the opening of a spatious view over the world: that was the moment of glory. But to toil upward, feel the gradient slacken and the top approach, as one does at the end of the Etchachan ascent, and then find no spaciousness of reward, but an interior – that astounded me. And what an interior! The boulder-strewn

> plain, the silent shining loch, the black overhang of its precipice, the drop to Loch Avon and the soaring barricade of Cairn Gorm beyond, and on every side . . . towering mountain walls.[43]

The poet's gaze looks *into* and *through* the mountain. As Robert Macfarlane notes in his prologue to a recent edition of the book, Shepherd provides a powerful corrective to our contemporary sensorial disengagement from nature.

> More and more of us come increasingly to forget that our minds are shaped by the bodily experience of being in the world – its spaces, textures, sounds, smells and habits – as well as by genetic traits we inherit and ideologies we absorb. We are literally losing touch, becoming disembodied.[44]

By reactivating this link and navigating us through her 'living mountains', Shepherd reminds us of the creative potential of the eye, rather than of its ordering powers. 'The eye brings infinity into my vision . . . It is the eye that discovers the mystery of light . . . the endless changes the earth itself undergoes under changing lights.'[45]

5 Mountains and Time

> There is a Clock ringing deep inside a mountain. It is a huge Clock, hundreds of feet tall, designed to tick for 10,000 years. Every once in a while the bells of this buried Clock play a melody. Each time the chimes ring, it's a melody the Clock has never played before. The Clock's chimes have been programmed not to repeat themselves for 10,000 years. Most times the Clock rings when a visitor has wound it, but the Clock hoards energy from a different source and occasionally it will ring itself when no one is around to hear it. It's anyone's guess how many beautiful songs will never be heard over the Clock's ten-millennial lifespan.[1]

The Clock of the Long Now is being built inside a limestone mountain near Van Horn, in western Texas. Its inventor, engineer and computer scientist Danny Hillis, conceived the project as 'a corrective' to Western contemporary societies' short attention spans. The acceleration of technology, the short-sighted perspective of market-driven economies or the next-election perspective of democracy all encourage us to focus on the near future. Building a clock that 'ticks every year' with the cuckoo coming out every millennium will, Hillis believes, prompt people to think in terms of generations and millennia, rather than years, and thus promote long-term responsibility.[2]

A site for a second clock has been recently purchased on top of another mountain in eastern Nevada. Both sites will be open to visitors, though they will be remote enough to generate an almost 'sacred' experience. 'To see the Clock you need to start at dawn, like any pilgrimage', says the project's official website. 'Once you arrive at its hidden entrance in an opening in the rock face, you will find a jade door rimmed in stainless steel, and then a second steel door beyond it.'[3] The mountain setting is as significant as the clock trapped in its rock. For every mountain resounds with the sound of time. Every mountain enshrines a mystery.

Beneath their eternal appearance mountains hide secret stories of ancient cataclysms and of slow perpetual transformation. Their mighty masses feature prominently in most accounts of creation. Hesiod (*c.* 700 BC) lists them, alongside the ocean, as the first things generated from the union between heaven and earth. According to Vedic cosmogonies, mountains, oceans and rivers remained enclosed, in embryonic form, within a golden cosmic egg created by Brahman (divine essence). 'After a thousand years Brahma made them manifest.'[4] In the Old Testament, after the giant flood Noah's Ark came to rest on top of Ararat. In the Quran the Prophet declares that the earth was created in two days and the mountains were then 'firmly' placed on it.[5]

Not only have mountain encounters contributed to ways of seeing and perceiving space, but they have shaped modern perceptions of time and history. Mountains' solid rock suggests immortality and immutability. It seems to elude the passage of

Building of the Clock of the Long Now in the Sierra Diablo Mountains, Texas.

time. We tend to forget that mountains' crisp contours have changed and will continue to change, just at a different timescale from humans.

According to John Ruskin, mountains moved. However, it was not until relatively recently that they had acquired their fourth, temporal dimension. Until the end of the seventeenth century, mountains were commonly deemed to have always been part of the same landscape God produced at the time of Creation. And it was not until the late nineteenth century that the idea that mountains slowly transform over time through erosive processes started to gain currency. Why was this the case? Why did mountains start to move? How have mountains contributed to a change in the way that we relate to our past – and to our future?

Biblical time

It all began with a Grand Tour. In 1671 Reverend Thomas Burnet, chaplain to William III, was asked to accompany the young Earl of Wiltshire to Italy to perfect his classical education. Yet for the clergyman the highlight of the trip was not the magnificent ruins of Rome, but the Alps. The scenery he encountered while crossing the Simplon Pass deeply shocked him. It was like nothing he had ever seen before, nor anything he could ever have imagined. Except for Philipp Clüver, Burnet lamented, mapmakers tended to downplay the representation of mountains and highlight instead human settlements, rivers and other features useful to 'Civil Affairs and Commerce'.[6] Sometimes peaks featured as small regular pyramids, at other times as elegant molehills or neat and ordered ranges. Such images were deceitful, as they conveyed a much more regular picture of the terrestrial surface than reality. Faced in person, the Alps were quite another thing:

> Suppose a man was carried asleep out of a plain Country amongst the Alps, and left there upon the top of the highest Mountains, when he wak'd and look'd about him,

> he wou'd think himself in an enchanted Country, or carried into another World . . . To see on every Hand of him a multitude of vast Bodies thrown together in Confusion, as those Mountains are; Rock standing naked round about him; and the hollow Valleys gaping under him; and at his Feet, it may be, an Heap of frozen Snow in the midst of Summer. He would hear the Thunder come from below, and see the black Clouds hanging beneath him; upon such a Prospect it would not be easy to him to persuade himself that he was still upon the same Earth.[7]

What Burnet saw was certainly more similar to a painting by the seventeenth-century Italian Baroque painter Salvator Rosa than to the maps he was used to. His own reaction was thus akin

Willem Janszoon Blaeu's map of Switzerland, *c.* 1640, engraving.

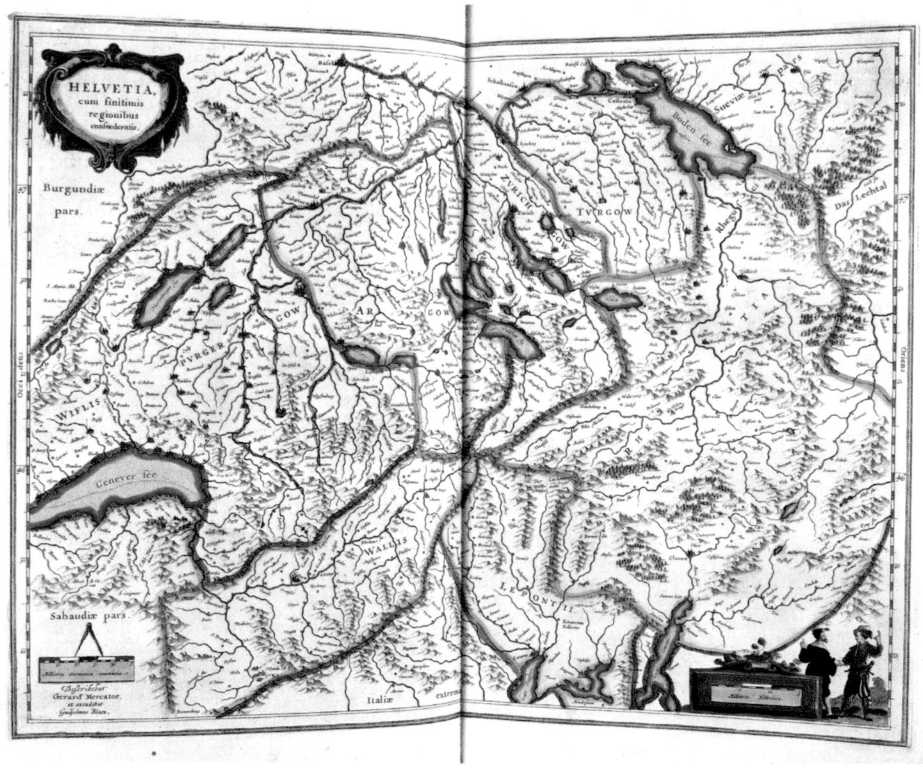

Diego Velázquez, *The Immaculate Conception*, *c.* 1618, oil on canvas.

to that of the 'sleeping man', that is, of someone totally unprepared to encounter such a scene.

What troubled the clergyman, however, were not so much the dimensions of the peaks as their disconcerting 'state of confusion'. Burnet was a son of his time and, as such, his aesthetic canons rested on symmetry, balance, containment – on the classical ideal of beauty. In Europe painters of his generation associated the Virgin's Immaculate Conception with a perfectly smooth incandescent moon and they were seriously disturbed by Galileo's discovery of the rough topography of the latter. The presence of mountains on the moon was interpreted by some as a symbol of a universal dissolution, for it meant that the moon was not the perfect ethereal object it had been believed to be, but a corruptible, mundane object, akin to the earth.[8]

Burnet's shock was even bigger. Those 'vast, wild and indigested heaps of stone and rubbish', as he called mountains, had 'neither Form nor Beauty, nor Shape, nor Order'. Indeed, he concluded, 'there is nothing in Nature more shapeless and ill-figur'd than an old Rock or a Mountain . . . They are the greatest example of Confusion that we know in Nature; no Tempest or Earthquake puts Things into more disorder.'[9]

The shock of the encounter forced the clergyman to interrogate his own beliefs on the origins of these features. How was it that the Creator in His infinite wisdom and mercy had produced such ill-figured formations? Where did their irregular shapes come from? For years the rocks continued to haunt Burnet. Having read Genesis and its commentaries again and again, the clergyman eventually came to the conclusion that mountains did not exist until after the Flood. The earth God initially created was smooth and beautiful – a 'cosmic Egg'. But the giant flood, a general confusion in nature, forever changed its shape. Mountains were therefore dramatic remnants and reminders of human sin – the same that caused God's wrath and the Flood.

'We still have the broken Materials of that first World, and walk upon its Ruins', wrote Burnet to the King. 'While it stood, there was the Seat of Paradise, and the Scenes of the Golden Age; when it fell, it made the Deluge; and this unshapen Earth we now inhabit, is the Form it was found in when the Waters had retir'd, and the Dry Land appear'd.' These ruins held the key to the mystery of Creation; they enabled Burnet to retrieve 'a World that had been lost for some thousands of Years, out of the Memory of Man, and the Records of Time'.[10]

Mountains also held the key to an apocalyptic future. With the Second Coming of Christ the earth would undergo a second global catastrophe: it would be consumed by fire and return to its original state. Christ and his resurrected saints would reign for a thousand years, after which the righteous would ascend to heaven, and the earth, no longer needed for human habitation, would become a star. The story is neatly summarized on the frontispiece to the second edition of Burnet's *Sacred Theory of the Earth* (1691), the most widely read geological treatise in the

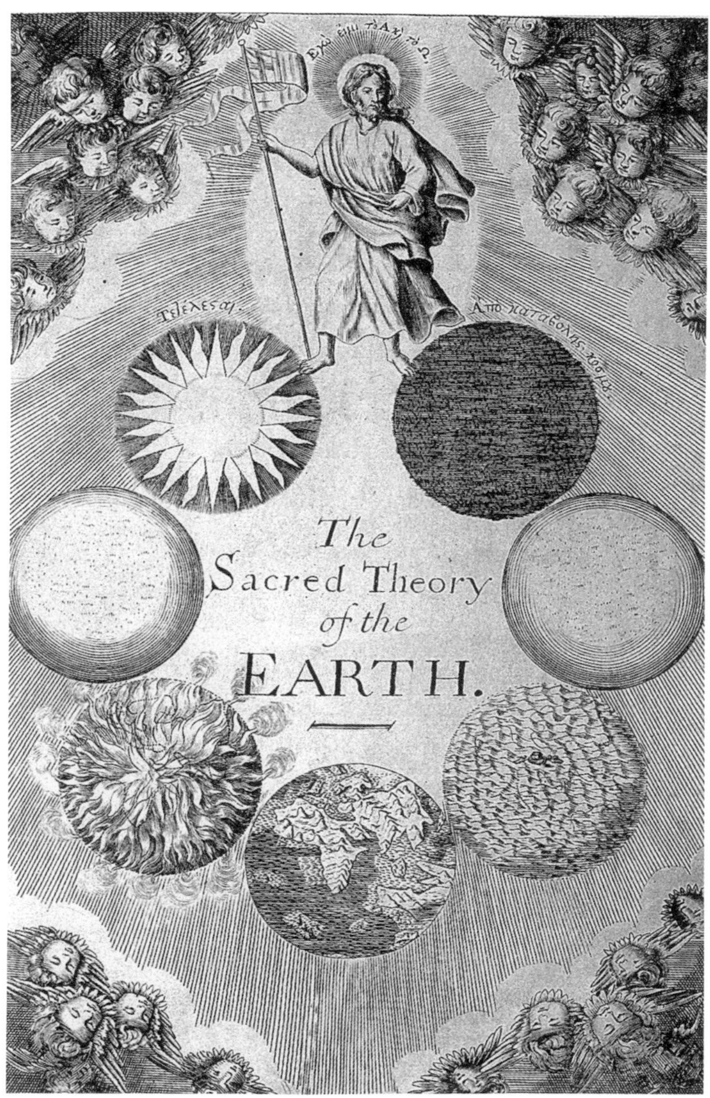

Thomas Burnet, *The Sacred Theory of the Earth* (1691).

seventeenth century and, as Macfarlane suggested, 'the first to have given mountains a past'.[11]

The image features the different 'states' of the globe in a circular clockwise chronology, from its messy creation to its end, or *telos*. The Greek word *telos*, however, embeds a second meaning: perfection. And perfection is reflected on both the smooth cosmic egg and the circular movement of time. 'I am the Alpha and the Omega' (9 Revelation 1:8, 1:11, 21:6, 22:13), says Christ, as he triumphantly towers over the first and last stages of Creation and winged heads of angels on the four corners of the image remind readers of humanity's apocalyptic future (as they did on medieval *mappae mundi* and on Renaissance Ptolemaic maps), after John's Revelation (7:1): 'I saw four angels standing on the four corners of the earth, holding the four winds of the earth'. As Burnet reminds the reader,

> To describe in like manner the Changes and Revolutions of Nature that are to come, and see through all succeeding Ages, will require a steady and attentive Eye, and a Retreat from the Noise of the World; especially so to connect the Parts and present them all under one View, that we may see, as in a Mirror, the several Faces of Nature from first to last, throughout all the Circle of Successions.[12]

Deep time

Burnet's time was the time of sacred history. It was simultaneously a time signposted by a linear sequence of events and a time of eternal returns – 'I am the Alpha and the Omega.' This time stretched no further back than 6,000 years, the age of God's Creation. It was not until the end of the seventeenth century that such temporal boundaries started to be severed. Hence, when the occasional traveller or natural scientist stumbled upon fossils while crossing the Alps, he would commonly regard the strange objects as material witnesses of the first stages of Creation or of the giant flood, which had once covered those summits.

The German philosopher and mathematician Gottfried Wilhelm Leibniz believed that the earth was a glass block, the product of a fusion that happened on the first day of Creation, when light was separated from darkness. His travels to the Harz Mountains and to the Alps and the observation of their irregular forms and matter convinced the philosopher that glass was the foundation of the earth, even if hidden by other bodies and materials. The upper parts of the vitrified crust had been shaped by hydraulic action. When the waters of the giant flood (and other floods) retreated, Leibniz believed, they returned to subterranean caverns, depositing fossils – which explained the presence of fossils of seashells and other marine creatures in the mountains.[13]

In the mid-eighteenth century Voltaire, the sceptic, challenged biblical interpretations – or at least, he tried. Fossils, he argued, were leftovers from 'picnics' of medieval pilgrims and crusaders traversing the Alps. 'Rotten fish were thrown away by a traveler and were petrified thereafter.' He was not taken seriously. 'The fossils present in the Alps are not the tastiest varieties of fish and would make a poor lunch,' sarcastically replied Georges-Louis Leclerc, Comte de Buffon, scorning the philosopher as a clown. In order to understand natural history, Buffon suggested, one had to dig much deeper 'into the archives of the Earth'.[14]

While Leibniz's and Voltaire's theories remained confined within the reassuring limits of traditional biblical chronology, Buffon interpreted each of the seven days of biblical Creation as metaphors for longer time spans. He thus publicly extended the age of the earth to 75,000 years, though he might have personally speculated about much longer periods.[15] Buffon was nevertheless just the precursor to a much more dramatic revolution: the opening of 'deep time'.

The invention of deep time is generally attributed to the Scot James Hutton, known also as 'the father of geology'. Besides being a naturalist, physician, chemical manufacturer and experimental agriculturalist, Hutton was also an indefatigable walker and a very careful observer. In 1785, during his wanderings

James Hutton, *Arthur's Seat and Salisbury Cross, Edinburgh*, replica of watercolour print.

through the Cairngorms, he found fingers of granite penetrating sedimentary rocks in a way such as to suggest that the former had been molten and forced themselves into the older rocks from below. This and similar anomalies he later observed in other parts of the country suggested a geological pattern of development and decay, of renewal and dessication. Rocks were eroded by external agents such as weather and at the same time 'coined in fire' and uplifted by heat and pressure. These processes, Hutton concluded, were part of an ongoing system that recycled the materials of the earth again and again.

This view, known also as Plutonism after the ancient Roman god of fire, opposed Neptunism (from the divinity of the sea), a popular theory promoted by Abraham Gottlob Werner, a professor at the Freiberg Mining Academy. According to Werner, the earth had originally consisted of water and the materials contained therein had sedimented over time, forming the core of the planet and the continents as a series of layers characterized by different types of rock. The giant flood repeated the process, adding further strata of rock and fossils.

Unlike Warner, comfortable within the reassuring boundaries of biblical time, the geological processes described by Hutton demanded spans of time unimaginable by the human mind – not hundreds or thousands, but *millions* of years:

> Time, which measures everything in our idea, and is often deficient to our schemes, is to nature endless and as nothing; it cannot limit that by which alone it had existence; and, as the natural course of time, which to us seems infinite, cannot be bounded by any operation that may have an end, the progress of things upon this globe, that is, the course of nature, cannot be limited by time, which must proceed in a continual succession.[16]

Hutton's theories opened up a new terrifying dimension, a 'temporal sublime' comparable to the 'opening of space' and the post-Newtonian aesthetics of the infinite of the previous century. They seemed to sink past and future into a bottomless chasm with, as Hutton wrote, 'no vestige of a beginning, no prospect of an end'.[17] Like mountaineers standing on the edges of a deep ravine or on the top of a lofty summit, Hutton's followers experienced a new type of excitement and vertigo. As his colleague and proponent John Playfair famously wrote of a visit to a geological site with Hutton, 'the mind seemed to grow giddy by looking so far into the abyss of time'.[18] Under the scrutiny of the geologist, mountains turned 'upside down'.

Geology boosted mountaineering, alongside an unprecedented interest in the earth's origins. 'Enlightened' late eighteenth-century naturalists such as Horace-Bénédict de Saussure were drawn to the Alps hoping to unveil the secrets of the past of the planet. Venturing to the mountains was no simple travel to 'a different world', as Burnet had argued. It literally meant embarking on a journey back in time. As one moved through different layers, whole eras passed by. 'Traversing the mountain,' wrote Louis Ramond de Carbonnières in his *Voyage to Monte Perdido and the Neighbouring Parts of the High Pyrenees* (1797), 'one travels from life to death.'[19]

By the early decades of the nineteenth century, landscape, and especially mountains, could no longer be looked at with the same eyes as before. In his 1817 cross-sections, the English surveyor William 'Strata' Smith dissected his own country. Villages and cities (including London) nearly disappear in between

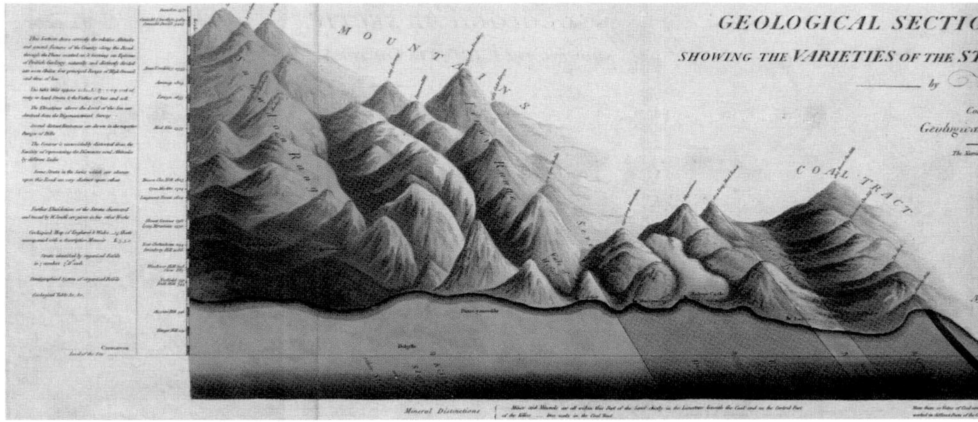

massive peaks, hills and the profundities of geological strata. Reduced to diminutive symbols, human settlements are swallowed by the magnitude of the earth and of time.

Likewise, in Caspar David Friedrich's *The Watzmann* (1825), a large painting inspired by Werner's theories and classifications of rocks, landscape features as a sequence of geological layers, each set on a different visual plane and devoid of human presence. The viewer's gaze travels from the irregular rock formations in the foreground to the gentle hills behind them and is further pushed towards the lofty snow-clad peak of the Watzmann towering in the background in all its grandeur. The mountain is no longer envisaged as an agent to human needs; as the art historian Timothy Mitchell has commented, 'it is greater than Man's fleeting passions and more enduring than Man's temporary structures.'[20]

Geological time nevertheless entered the popular Western imagination mainly through the work of Charles Lyell. In 1828 the Scottish geologist had travelled to Sicily to study Mount Etna. In between the lava layers, he observed thick strata of oyster shells, an indication that the time between lava flows must have been significant. The volcano exceeded 3,000 m. Lyell thus concluded that such a massive cone had been built through aeons of small eruptions.[21] As with Hutton, Lyell portrayed a planet whose age was in millions of years. However, while for

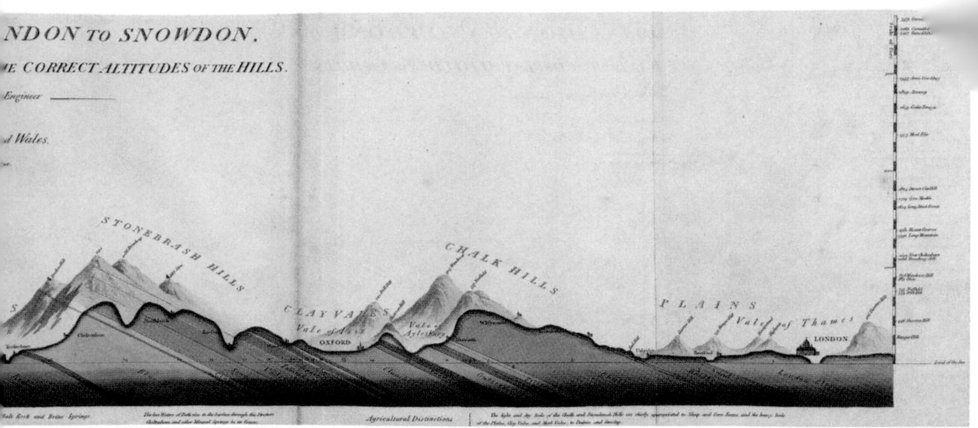

William Smith, *London to Swindon geological cross-section*, 1817.

Hutton periods of uplift could be global and catastrophic, for his successor all the stages in his cycle operated locally and simultaneously, giving the earth 'a timeless steadiness through all its dynamic churning'.[22]

Lyell's *Principles of Geology*, published two years after his visit to Etna, soon became a best-seller. Scientific observations intertwined with vivid descriptions of unfamiliar sceneries captivated the attention of thousands of readers. Victorian searchers of the picturesque were offered tours of Herculaneum and Pompeii at the foot of Vesuvius; they were taken to the volcanoes of Greece, Iceland and Mexico; they were lifted to the top of Etna and Mt Teide and transported down to the mysterious worlds underneath the earth's surface to explore their slowest motions.

Under Lyell's gaze, even the stillest landscape assumed a new, dynamic, temporal dimension. In describing Etna's great plain of Val del Bove, the British geologist invited his readers to imagine themselves in a large amphitheatre surrounded by deep precipices broken by vertical walls of lava. In autumn, observed Lyell, 'their black outline may often be seen relieved by clouds of fleecy vapor which settle behind them, and do not disperse until mid-day, continuing to fill the valley while the sun is shining on every other part of Sicily, and on the higher regions of Etna.'[23] Armchair travelling with Lyell meant embarking on a process of continuous revelation:

Caspar David Friedrich, *The Watzmann*, 1825, oil on canvas.

> As soon as the vapors begin to rise, the changes of scene are varied in the highest degree, different rocks being unveiled and hidden by turns, and the summit of Etna often breaking through the clouds for a moment with its dazzling snows, and being then as suddenly withdrawn from the view. An unusual silence prevails; for there are no torrents dashing from the rocks, nor any movement of running water in this valley such as may almost invariably be heard in mountainous regions.[24]

General interest in the earth's deep past was further stimulated by Louis Agassiz's speculations, a decade after the publication of Lyell's *Principles,* about the existence of a remote ice age. The Swiss geologist connected the effects of this era to the contemporary glaciers of his mountains, 'still moving, still scouring rock and depositing great boulders far from their place of formation'. Glaciers and traces of former glaciers thus came to be interpreted as the most dramatic vestiges of this deep past, as giant 'scrolls of time' upon which the story of the planet had been written, layer upon layer.[25]

Agassiz's ideas had a global reach. By the early 1870s John Muir had come to believe that Yosemite had been carved out by glaciers and Eadweard Muybridge, the pioneer of motion picture projection, travelled there to photograph glacial traces in the landscape. As the writer Rebecca Solnit comments, 'in Yosemite, water and rock became Muybridge's principal subjects. The water spoke of change, of the passing moment, and the rock of what endures, of geological eons'.[26] In a sense, the two elements became metaphors for the different temporal dimensions between which Muybridge was caught and which he tried to control through photography: on the one hand, a time shrunk by new communication and transportation technologies, such as the telegraph and the train; on the other, a time endlessly expanded through the slowness of geological processes.

No. 22.
View from the summit of Etna into the Val del Bove.
The small cone and crater immediately below were among those formed during the eruptions of 1810 *and* 1811.

Charles Lyell, View from the Summit of Etna into the Val del Bove, woodcut from *Principles of Geology* (1833).

Besides traditional photographs, Muybridge also produced pictures of Yosemite to be viewed through the stereoscope, an optical device for viewing pairs of photographs as a single three-dimensional image. The stereoscope emerged from research into binocular vision in the 1820s and '30s and was commercialized as a visual attraction in the 1850s. In many ways, it complemented panoramas, telescopes and the other optical devices that were so popular with the Victorians. As with panoramas, one of its most popular uses in the second half of the century was as a medium for armchair travel, reproducing 'surrogates' of distant places.

By the end of the century, besides Muybridge's popular Yosemite series, stereocards featuring the desolate rocky landscapes of the American West, with their bare cliffs, isolated mesas and buttes, had become standard items in Victorian collections. The stereoscopic medium simultaneously conveyed a sense of rocky solidity and uncanny displacement – as if time had suddenly frozen. These cards featured alongside others depicting picturesque ruins: abandoned gothic churches, moss-covered Greek temples, solitary broken pillars facing the sea.

Mountains and ruins had indeed become intimately connected. In the last quarter of the century American geologists

Eadweard Muybridge, *Falls of the Yosemite, Second Fall at low water*, 1868.

came to regard mountains as vestiges of their former selves. While the slow and uniform geological processes described by Lyell were primarily tectonic and depositional, before the late nineteenth century decay was usually not even considered. Part of the reason is that the verdant Alps and the other European mountains studied by Lyell and his predecessors bore scarce direct resemblance to ruins. By contrast, the isolated mesas of the American West recalled their erosional origins 'with diagrammatic clarity'.[27]

Thus, wrote John Wesley Powell in 1875, 'The mountains were not thrust up as peaks, but a great block was slowly lifted, and from this the mountains were carved by the clouds – patient artists who take what time might be necessary for their work.'[28] Geological time no longer caused only curiosity or vertigo, but an enduring sentiment of melancholy akin to the contemplation of ruins and best expressed through the pen of the poet. Under the starry sky, in the utter silence of the lonely mesas, wrote John Gould Fletcher,

> . . . all things are crumbling slowly,
> The stone, the dream
> And the effort . . .[29]

Classical time

Nineteenth-century interest in the earth's deep past was paralleled by a renewed interest in the more familiar historical past, and especially in the classical past. Sublime time coexisted with the time of the beautiful. When related to human history, however, mountains were usually seen less as crumbling ruins or changing presences in the landscape than as eternal monuments and immutable landmarks for orientation within this past: 'the great landmarks of human history', wrote Joel Tyler Headley.[30]

In 1813 Edward Daniel Clarke, a classicist and mineralogist at Cambridge who had been a tutor for several distinguished families and had already made himself famous for travelling from Scandinavia to Syria and plundering a quantity of antiquities in Ottoman Greece and Egypt, climbed several 'mythical peaks', including Mount Gargarus in Asia Minor (1,767 m). In ancient times this mountain was dedicated to the worship of the goddess Cybele and is the highest point of the Ida range in Turkey, southeast of the location of ancient Troy.

Clarke's ascent activated a temporal descent from the present to the past. The British clergyman and his retinue first traversed a zone characterized by a scenery resembling 'the country in the neighbourhood of Vietri, upon the Gulf of Salerno, where Salvator Rosa studied and painted the savage and uncouth features of Nature, in his great and noble style'.[31] The party subsequently passed by 'the remains of some small Greek chapels, the oratories of ascetics, whom the dark spirit of superstition, in the fourth century of the Christian era, conducted from the duties of civil society, to the wildest and most untrodden solitude'.[32] Finally, on the top of the mountain, Clarke gained a bird's-eye view of the lands and landmarks of the classical world – a totalizing, ordered view akin to Saussure's:

> What a spectacle! All European Turkey, and the whole of Asia Minor, seemed, as it were modelled before me on a vast surface of glass . . . Looking down upon Troas, it appeared spread as a lawn before me. I distinctly saw the course of the Scamander through the Trojan Plain to the sea. The visible appearance of the river, like a silver thread, offered a clue to other objects. I could discern the Tomb of Aesyetes.[33]

Clarke's experience and enthusiasm were by no means unique. By the time the clergyman was standing on the summit of Gargarus, climbing the famed peaks of antiquity had started to become a fashionable component of Grand Tours. For example, before heading to Mont Blanc in 1741, Richard Pococke had just climbed Mount Athos, Ida, Vesuvius and the pyramids of Giza, whereas in 1809 Lord Byron ventured to the slopes of Parnassus in search of poetic inspiration. The encounter with the *real* mount, sacred to Apollo and the Muses, he argued, gave him a more vital sense of the poetry it inspired than all his previous classical readings.[34]

For centuries ancient Greeks and Romans had been venerated as the ancestors of the modern Europeans. Yet knowledge of the ancient world remained predominantly literary and irremediably mediated through Rome. On their Grand Tours, Western scholars like Burnet travelled to Italy but did not dare cross the Ionian Sea. Only occasional lunatics would venture to the then brigand-infested lands of Homer and Pericles. It was not until the turn of the eighteenth century that Grand Tourists started to turn their gaze to the East. The Napoleonic Wars caused them to divert their routes from Italy and Central Europe towards the Ottoman Empire.

At the same time, 'the romantic idealization of the ancient Greeks, [linked to] an enthusiasm for freeing modern Greeks from the Turk yoke', inflamed the imagination of European scholars with a novel force.[35] From 'ancestral European myth', the classical past literally became a 'foreign country'. Time was turned into space. 'Beyond the Ionian Sea', wrote Edward Dodwell in the early 1800s, 'almost every rock, every promontory, every view is haunted by the shadows of the mighty dead. Every portion of the [Ottoman Greek] soil appears to teem with historical recollections.'[36]

Armed with Pausanias and Homer, professional historical topographers, diplomats, clergymen, naval officers and other zealous philhellenes embarked on careful surveys of their idealized literary past in the physical landscape. As with Saussure, Carbonnières, Hutton, Lyell and their followers, theirs were journeys to the past. Yet, unlike those of the early geologists, these were less journeys of discovery than of reconnaissance – surveys of a past the men felt intimately familiar with, even before leaving home. Names like Olympus, Ida, Ossa, Pelion and Athos had signposted the readings of their youth.

Greek mountains bounded the visual and imaginative horizon of this poetic past, in the same way that they appeared on travellers' drawings and paintings, from Edward Dodwell and Charles Robert Cockerell's erudite topographies to Edward

Edward Lear, *Mount Parnassus with a Group of Travellers*, 1879, watercolour.

Lear's open vistas. 'With all its drawbacks, [the] wild life [of the Levant] has great charms,' wrote George Ferguson Bowen, Secretary of Government in the Ionian Islands and president of the University of Corfu from 1847 to 1851.

> The first rays of the sun gilding the summit of Athos, or Olympus, or Pentelicus, or Parnassus, or Ida, or Lebanon, or of some other mountain of many memories which is sure to bound your horizon in the East, place you on the saddle, after a refreshing swim in the Aegean, or a plunge in some classic stream, if the sea be too far off; and the first pale beams of the rising moon or of the evening star bid you sink, like a bird of the forest, to rest.[37]

Unlike the Western Alps, Hellas' peaks were not perceived as sublime or chaotic, but rather as inspiring poetic objects, as beautiful, well-defined landmarks for orientation (as they had once been to the ancient sailors and heroes they read about), and as privileged nodes within a complex web of memory.

Greek mountains also served as exceptional platforms from which to observe ancient landscapes and reactivate classical memory in the orderly fashion Burnet would have perhaps expected. 'When standing on the summit of Parnassus, which commands the most extensive view of Greece, reaching from the north of Thessaly to Arcadia, and from the entrance of the Corinthian Gulf to the extremity of Attica', wrote the writer and traveller Henry Fanshawe Tozer in 1882,

> most of these summits are visible, and the effect produced is – not as in looking from Etna over Sicily, where everything is so dwarfed below you as to resemble an outspread map, nor yet in some Alpine views, where the attention is absorbed by one overpowering object – but that the eye passes on from point to point, and rests equally on one after another of this federation of mountains.[38]

Holy nations

Much of the increasing popularity of Grand Tours (and mountaineering) in the Levant had to do with the establishment of a Kingdom of Greece in 1832 backed by the European powers and with its progressive expansion. As well as a German sovereign, the new kingdom inherited Western aesthetic sensibilities, including the Romantic fascination with the sublime and the picturesque – and thus with mountains. In the newborn kingdom integration was achieved textually before it was on the ground. Hellas' 'mutilated body' was restored through the natural settings of Romantic poems, novels and paintings. For the Greeks and their admirers mountains thus became timeless 'monuments of the nation', grounding national narratives and commemorating heroic feats.[39]

Panagiotis Soutsos' *The Wayfarer* (1831), a tragedy in five acts and one of the most emblematic products of Greek Romanticism, is dominated by a continuous intertwining between the inner drama of the protagonist, persecuted by the ghost of his abandoned beloved, and sublime natural settings. Soutsos' pen gives life to inert landscapes; it transforms 'natural monuments' into monuments of the new nation. Lost in the ecstatic contemplation of Mount Athos (where most of the tragedy takes place), the Wayfarer weaves one of the most impressive hymns of modern Greek literature. The mountain assumes anthropomorphic traits.

From the height of his immortality, the petrified giant observes the passing of time and the caducity of human life; he takes the word; he breathes. Within the colossus' body, Aristotelian elements mix in all their drama. Rock and vegetation, torrents and waves, thunder and lightning fuse into a creature at once eternal and pulsating with life. Athos becomes an emblem of Hellas' body politic – at once a timeless monument and a dynamic creature. It becomes a material link between human transience and the immortality of a newborn nation-state, as well as between Greek religious sentiment and European Romantic sensibility.[40]

On nineteenth-century paintings, Greek pyramidal peaks loom on the horizon of iconic landscapes scattered with ruins, as to echo the immortality of the nation and its heroes. For example, in Joseph-Denis Dionysius Odevaere's *Lord Byron on His Death Bed* (*c.* 1826), such a landscape is theatrically unveiled behind the dying body of the Apollonian poet-hero. Framed by a purple curtain and a statue of Eleutheria (Liberty), this landscape counterbalances Byron's broken lyre, suggesting that the ideals and nation he fought for and that he celebrated through his muse shall outlive his premature death.

Modern Greece is but one of several instances in which mountains have been used as eternal monuments of the nation. Nearly a century before Greece's independence, the Swiss Alps

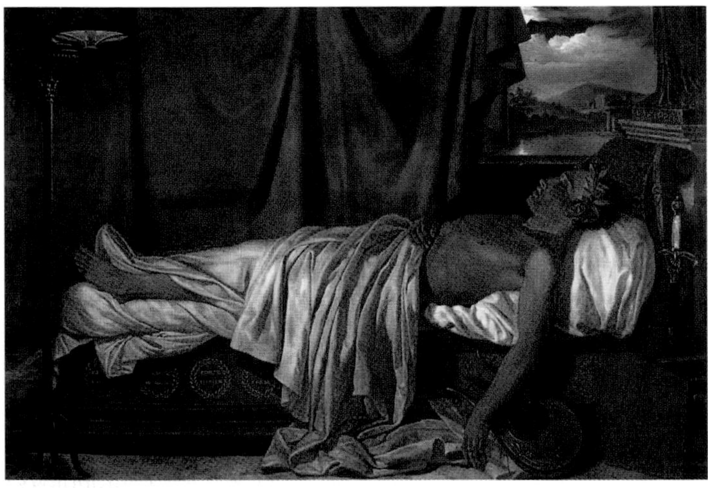

Joseph-Denis Odevaere, *Lord Byron on His Death Bed*, *c.* 1826, oil on canvas.

Albert Bierstadt, *Storm in the Rocky Mountains, Mount Rosalie*, 1866, oil on canvas.

had become 'seats of virtue' and synonymous with freedom, largely thanks to the writings of Albrecht von Haller, Rousseau and others. In other national histories mountains are tangible signs of the nation's manifest destiny, whether as natural barriers preserving ancient traditions and safeguarding national unity (or unification, as in the case of nineteenth-century Italy and Bohemia), as natural backbones (in Korea), as bedrocks for independence (in Norway from Sweden or in the case of Catalonia from Spain) or simply as majestic eternal symbols (for example, in the United States).

As opposed to the ruin-scattered nineteenth-century Greek landscape paintings and their moderate poetic mountains, in Albert Bierstadt's paintings or Muybridge's photographs of Yosemite human features tend to disappear altogether, or become insignificant presences used by the artist to convey the scale of nature's grandeur. 'My first view of the Rocky Mountains', recalled Bierstadt, 'had no way of expressing itself save in tears. To see what they looked like, and to know what they were, was like a sudden revelation of the truth, that the spiritual is only real and substantive; that the eternal things of the universe are they which afar off seem dim and distant.'[41] America's chronology was not the chronology of classical European civilization; it was the timescale of nature inherited directly from the Creator and visually translated through the vastness of its pristine spaces, through the magnitude of its mountains and giant sequoias. In the new continent time became a space for spiritual redemption.

The Magic Mountain

The idealization of mountains as ancestral spaces and sites of virtue implies a sense of timelessness. This is a theme that is central to Thomas Mann's *The Magic Mountain*, a novel written at the outset of the First World War and published in 1924. Yet there are different ways of approaching timelessness, just as there are many ways of seeing mountains. While the mountains in the national narratives mentioned above are approached from ground level as immutable objects, in *The Magic Mountain* perspective shifts. The reader is elevated from the ground to a mountain's slopes. As a result of this spatial shift, time too shifts, from the collective to the introspective time of the sacred mountain. The mountain becomes a still platform from which to put the mobile world beneath into perspective.

Like Hemingway's 'The Snows of Kilimanjaro', *The Magic Mountain* was inspired by a real-life experience. In 1912 Mann's wife, who was suffering from a lung complaint, was confined for several months in a sanatorium in Davos, high up in the Swiss Alps. During his protracted visit to her, Mann became acquainted with the cosmopolitan institution and its staff. Like Mann, the protagonist of the novel, Hans Castorp, undertakes a journey to see his tubercular cousin at the Davos sanatorium. Away from the world of 'the flatlands' and its obligations, he finds himself immersed in the rarefied atmosphere of a microcosm in which time seems to be suspended between life and death. With its international clientele, the Swiss sanatorium is 'a model of a decadent Europe on the Eve of World War I – a world that, for all its affluence and sophistication, is drenched in the sweet-sickly odour of death. And how can it be otherwise, so far is it removed from the active life of normal people on the plains below?'[42]

Days, weeks, months pass by. They indistinguishably flow into each other. Eventually Castorp himself develops symptoms of tuberculosis and is persuaded to remain in the sanatorium until recovery. He ends up staying there for seven years, during which, freed from time constraints, he is able to explore his inner self and the meaning of life. 'Can one tell – that is to say, narrate – time, time itself, as such, for its own sake?' asks Mann. The question is rhetorical, for, he then explains, 'that would surely be an absurd undertaking'. Storytelling and music can only present themselves as a flowing, as 'a succession in time, as one thing after the other'.[43] Ironically, the timeless mountain becomes a site of awakening for both protagonist and reader.

The novel concludes with the outbreak of the Great War, the greatest tragedy of modern European nations. Castorp returns to the flatlands and volunteers for the military – with a hope: 'Out of this universal feast of death, out of this extremity of fever, kindling the rain-washed evening sky to a fiery glow, may it be that Love one day shall mount?'[44]

Mountain histories swing between contingency and eternity, between the present and the deep time of their creation. No more manifest is this tension than in national histories and in their continuous attempts to anchor modern traditions to an ancestral land – and to ancestral ranges. Kingdoms, nations and states have used mountains to fix their 'natural' boundaries. However, mountains and nationalisms are also strongly intertwined at a deeper spiritual and symbolic level whereby life and death encounter each other. Nationalist sentiment is usually expressed as reverence towards a shared mythical past, a past that is 'naturally' embedded in the landscape. National histories imbue peaks and ranges with a feeling of the sacred – a feeling demanding an indefinite sacred time.[45] No matter how different their mountain topographies, what these stories share is a common belief in a golden age, a quest for purity, authenticity, virtue and regeneration – and the materialization of all this through the rock and the terse air of high places.

Yet if the nineteenth and twentieth centuries were centuries of nation-states and natural histories, we might wonder whether or to what extent this is still the case today. Just as globalization calls for new models and ways of imagining space, the anthropocene, the new 'man-made' geological epoch we are living in, invites us to think about time and history differently, or at least at a different scale. From eternal monuments or melancholic ruins, nowadays mountains, with their shrinking glaciers, have become delicate barometers of climate change. As with Hillis's 'Clock of the Long Now', they call us to reflect not only on the deep past but also on a 'deep future'.

6 Mountains, Science and Technology

Lofty peaks have been privileged observatories for the study of natural phenomena since antiquity. Strabo reports regular ascents of Etna undertaken for this specific purpose. Seneca commissioned his friend Lucilius, procurator of Sicily, to climb the volcano to detemine the veracity of reports that it was gradually sinking, for, he argued, 'At one time it used to be visible to mariners from a greater distance than at the present.' Pliny the Elder, who dramatically lost his life while attempting to examine the eruption of another volcano, Vesuvius, wrote about individuals who wandered across 'inaccessible mountain summits and remote wildernesses' in order to pursue research on plants.[1]

Some peaks were renowned throughout the ancient Mediterranean for specific features and phenomena associated with them. Mount Athos, for example, was famous for its mighty stature. According to Pomponius Mela, writing in the first century AD, the mountain reached the highest layer of the atmosphere where rains and clouds were formed. The idea, Pomponius argued, 'gets credibility because ashes do not wash off the altars that it has on its peak but remain on the mound where they are left'.[2] Mount Ida in Asia Minor was associated with another singular atmospheric phenomenon. In summertime, Diodorus Siculus wrote,

> on the summit of its peak, about the time of the rising of the dogstar, owing to the stillness of the surrounding atmosphere, the highest point is far above the current of

> the winds and while it is still night the sun is seen to rise, emitting its rays not in a spherical form, but so that its brilliancy is dispersed in various directions, with the appearance of a number of flames striking the horizon.[3]

In the fourth century, the heights of Cappadocia offered Church Fathers and hermits privileged platforms from which to contemplate creation and helped shape their understanding of the cosmos. From the height of his hut on a lofty peak in Peristrema, for example, St Basil the Great commanded the extended plain at the feet of the mountain. As vapours gradually dissipated, the hierarch observed the varieties of trees covering the slopes of the mountain; he followed the course of the river Iris, and, at night, lifted his eyes up to the vault of heaven and its stars. His mountain idyll acquainted Basil with the works of the Creator and their functioning, in which he saw tangible 'memorials of His wonders'.[4]

In modern Europe, mountains maintained and expanded their ancient role as observatories. Some of them became tied to important scientific experiments. For example, in 1774 Schiehallion, an isolated 1,000-m-high peak in central Scotland, was used to determine the mean density of the earth. As we have seen, Saussure's ascent of Mont Blanc was itself motivated by scientific experimentation. The most famous test, however, was conducted on the Puy-de-Dôme, a 1,464-m lava dome in Auvergne, south-central France.

In 1648 Pascal sent his brother-in-law to this peak to experiment with pressure. The French scientist-philosopher could not ascend the mountain himself as he was a semi-invalid, but he believed that anyone, if properly instructed, could perform the experiment. Since air weighs less at the top of a mountain than at its foot, Pascal speculated, the levels of mercury in the barometer should fall as there would be less air to hold it up. On Puy-de-Dôme Pascal's hypothesis was proved and the experiment was repeated on the Saint Jacques tower in Paris, with the same results.[5]

More significantly, changes in the level of mercury were used to determine altitude. In the eighteenth century mountain peaks thus started to be measured. The Italian scientist (and inventor of the battery) Alessandro Volta, who travelled to the Alps in 1777 and 1787 to conduct experiments with air and electricity, became one of their most enthusiastic measurers, though struck by the 'ruined appearance' and 'diabolic desolation of that inorganic realm'.[6] Throughout the following centuries mountains became objects for systematic scientific investigation. Modernity tamed high places. As observational platforms on the world, mountains nonetheless continued to maintain both their sacred aura and their demonic cast.

Chimborazo

'Taming mountains' did not simply mean establishing their exact altitude. It also meant naming them and, perhaps even more significantly, setting conceptual boundaries around them. By the early nineteenth century mountains were looked at not only as sublime sites of wonder, but as 'islands on the land' or vertical microcosms, characterized by different altitudinal microhabitats and therefore by different types of vegetation. For example, Johann Reinhold Forster, the German naturalist who accompanied Cook on his second voyage, was struck by the insular resemblance of Table Mountain at the Cape of Good Hope in South Africa 'in the sense of being a circumscribed physiographic form upon which processes could be observed operating'.[7]

The main individual responsible for this way of envisaging mountains in the modern Western scientific imagination is nevertheless Alexander von Humboldt. Between 1799 and 1804 the Prussian naturalist travelled extensively in South America carrying with him more than three dozen scientific instruments. Unlike Forster and other naturalists, who were after new floristic and animal species or minerals, Humboldt was not interested in the collection of data and specimens per se, but rather in the principles underpinning their distribution. More broadly, he was interested in the world as a whole, a 'great integrated whole, moved and animated by internal forces'.[8]

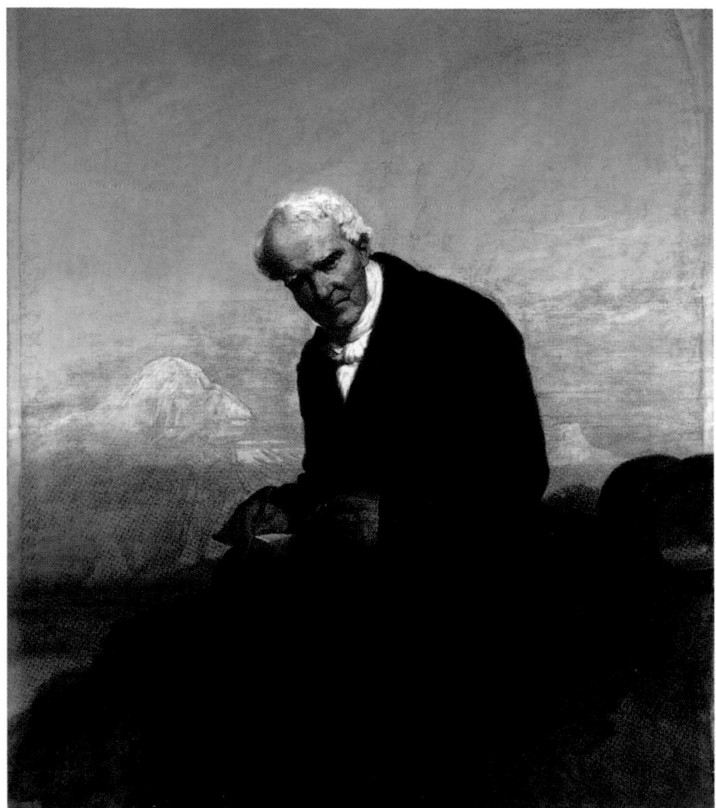

Julius Schrader, *Baron Alexander von Humboldt*, 1859, oil on canvas.

Humboldt laboured under the Goethian notion of harmony in nature and unity in diversity. Plants and organic forces, he believed, were distributed over the globe according to mathematically determinable limits and could thus be mapped in the same way as other phenomena, such as heat or magnetism, for example. He therefore privileged synthesis over isolated facts and, for example, criticized Linnaean botanists for exhausting their energies with 'details' while missing 'the big picture', occupied as they were with the discovery of new species and their classification. He stated that he 'would much rather know the exact and elevational limits of an already known species than discover fifteen new'.[9]

It was for this reason that Humboldt selected the South American interior for his observations. With their biotic and climatic diversity, the Andes offered the most suitable site for Humboldt's unitarian vision and his ideal of global knowledge. In the cordilleras the German naturalist envisaged a vast observatory enclosing the entire variety of the cosmos, an extraordinary space in which global order was made locally accessible.

As with Saussure on Mont Blanc, on Chimborazo, a 6,268-m inactive volcano and the highest peak in Ecuador, Humboldt found 'all the phenomena that the surface of our planet and the surrounding atmosphere present to the observer' and physically saw 'the general results of [his] five years in the tropics'.[10] The majestic mountain, which was then believed to be the highest in the world, condensed huge expanses of territory into a single vertical ascent. In a perpendicular rise of 4,800 m,

> the various climates succeed one another, layered one on top of the next like strata . . . There, at a single glance the eye surveys majestic palms, humid forests of *Bambusa*, and the varied species of *Musaceae*, while above these forms of tropical vegetation appear oaks, medlars, the sweet-brier, and umbelliferous plants, as in Europeans homes . . . There the different climates are arranged the one above the other, stage by stage, like the vegetable zones, whose succession

Alexander von Humboldt, *Physical Portrait of the Tropics*, 1807.

> they limit; and there the observer may readily trace the laws that regulate the diminution of heat, as they stand indelibly inscribed on the rocky walls and abrupt declivities of the Cordilleras.[11]

The Ecuadorean peak was employed by Humboldt in his famous diagram *Physical Portrait of the Tropics* (1807) to exemplify his vision. Situated between the depths of the terrestrial crust and the higher regions of the atmosphere, Chimborazo here features as an ordered microcosm and meeting point of elements, with its snow-capped summit, its luxuriant vegetation and a column of smoke quietly spewing from the chimney of Cotopaxi (the second highest peak in Ecuador). The earth's entire climatic and vegetational spectrum is vertically mapped on Chimborazo according to elevation above sea level. Other phenomena are likewise represented in relation to altitudinal change: the progressive diminishing of gravity and atmospheric pressure, the intensification of the blue colour of the sky, the weakening of light and so on. As beholders ascend the various layers, they are repeatedly reminded of the heights of more familiar peaks, including Mont Blanc.

Chimborazo reappeared in later representations, including a world map of vegetal distribution in Berghaus's *Physikalischer Schul-Atlas zum Kosmos von Alexander von Humboldt* (1850). This time the mountain is set side by side with more modest peaks situated at higher latitudes: Mont Blanc and Mont Perdu in the temperate zone and Sulitjelma (Lapland) in the frigid zone. Here the beholder's gaze is allowed to pierce the vegetal mantle and get a glimpse of the often complex geologies of the various peaks. At the height of the dispute between Neptunists and Plutonists over the birth of mountains, Humboldt turned his attention to the 'internal make-up' of our planet, and thus to the material composition of mountain peaks and the physics of the earth.

On this and other similar representations, Chimborazo features simultaneously as a summary and as part of the broader cosmos, which the naturalist carefully mapped. Repeated precise

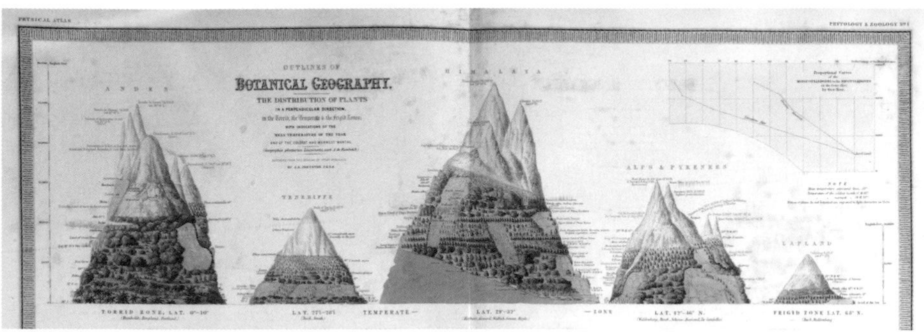

Detail from plant distribution map in Berghaus's *Physikalischer Schul-Atlas zum Kosmos von Alexander von Humboldt* (1850).

measurements and mapping were nevertheless just one way to engage with creation. Humboldt's approach was twofold. Besides attentive observation of the landscape physiognomy and the rational investigation of physical laws, the German naturalist encouraged poetic abandonment to the grandeur of nature. The sensual interfusion with the mountain and its elements, he argued, helped the scientist achieve an intimate, spiritual contact with the cosmos:

> When the human mind first attempts to subject to its control the world of physical phenomena, and strives by meditative contemplation to penetrate the rich luxuriance of living nature, and the mingled web of free and restricted natural forces, man feels himself raised to a height from whence, as he embraces the vast horizon, individual things blend together in varied groups, and appear as if shrouded in a vapory veil.[12]

Mountain landscapes allowed Humboldt to order nature at a glance, to mediate between the local scale and the cosmos, to grasp the hidden forces that animate the earth. 'Everywhere', he writes, 'the mind is penetrated by the same sense of the grandeur and vast expanse of nature, revealing to the soul, by a mysterious inspiration, the existence of laws that regulate the forces of the universe.' Yet it is not 'peaceful uniformity' but landscape's diversity and the diversity of landscapes that move the human heart. 'Every zone of vegetation', argues Humboldt, 'has, besides its own attractions, a peculiar character, which calls forth in us special impressions.'[13]

Nowhere are such impressions captured more powerfully than in the huge canvases by Frederic Edwin Church. Inspired by Humboldt's writings, the American artist travelled to South America between 1853 and 1857. Following the traces of the Prussian naturalist, Church sought to capture the dramatic grandeur of equatorial volcanoes and the lively variety of tropical landscapes. In *The Heart of the Andes*, a 1.7-by-3-m oil painting first exhibited in 1859, stunning botanical detail foregrounds a glistening pool served by a waterfall, while the snow-capped Chimborazo majestically towers in the distance behind intermediate dark slopes. Like Humboldt's maps and diagrams, Church's paintings were not snapshots, but composites of the topographies he observed during his travels. They encompassed an entire Andean landscape, 'from Arctic pinnacles, through temperate zones, to the palpable humidity of tropical valleys' – a range as grand as the paintings' sizes. As the geographer Edmunds Bunkse observes, 'There is certainly no one place where a person could behold at once this vast continental scene.'[14] In

Frederic Edwin Church, *The Heart of the Andes*, 1859, oil on canvas.

Frederic Edwin Church, *Tropical Scenery*, 1873, oil on canvas.

this sense, Church's paintings are hyperrealistic and yet at the same time imaginary panoramas.

Adopting and magnifying the conventions of the picturesque, Church plays with proximity and distance in *Tropical Scenery*. As light quietly filters through the layers of tropical humidity, the American painter juxtaposes fine detail to hazy slopes looming in the distance, empirical observation to imagination. Likewise, Humboldt's views from the Andes are always characterized by a haze on the horizon, by a progressive loss of clearness and transparency as distance increases. For Humboldt and Church, mountain landscapes are dynamic thresholds. They are spaces of possibility, meeting points between the present and what has yet to come. They mediate between the locality and the cosmos, thus allowing the human mind to grasp the hidden forces that animate the earth.

Faust

Humboldt's and Church's tropical mountain views embrace a dream of omniscience. In some respects, they are akin to the view Mephistopheles, Satan's envoy, presented to Goethe's Faust, the prophet of modern science and development.

Dr Faust withdrew from society and embarked on a solitary intellectual quest that in the end brought him only dissatisfaction and despair. Having traversed all the realms of human knowledge – from physics and medicine to theology – he eventually turned to black magic. This granted him access to the secrets of the cosmos, to 'the force that binds creation's inmost energies' and 'her vital powers' – the same force Humboldt sought in his Andean landscapes. For Faust, however, the encounter resolved not only in an epiphany, but in a transfiguration. 'The powers of Nature here, all around revealing. Am I God – so clear mine eyes?'[15]

The energies of nature are not external to the scientist. They animate his inner microcosm in the same way they animate the cosmos. Inside Faust is a volcano of repressed energies, an abyss of desires no less infinite than the misty horizons envisaged by Humboldt and Church. Mephisto promises Faust every mode of human experience, a process that will lead the man to an unending growth, to an explosion of creative and destructive powers beyond his control. This process, Marshall Berman observes, shall lead Faust to a threefold reinvention of himself as a visionary, lover and developer. Intriguingly, these metamorphoses are signposted by mountains.

In the first part of the tragedy, the lone scientist overlooks the simple medieval world he has left behind him from the height of his abode. Mephisto takes him down into that world to experience inebriation and love, and then up again on impossible flights. For Faust the vital thing is to keep moving – imaginatively, visually and physically.

> Then should I see the world below,
> Bathed in the deathless evening-beams,

The vales reposing, every height a-glow,
The silver brooklets meeting golden streams.
The savage mountain, with its cavern'd side,
Bars not my godlike progress.[16]

As the drama unfolds new peaks emerge and new horizons open up. Yet everything Faust touches he ends up destroying, including the innocent girl with whom he has fallen in love.

Faust's final metamorphosis occurs once again in the mountains, but this time at a moment of impasse. Having explored endless experiential possibilities, having travelled through all history and mythology, Faust now stands still on a fierce, jagged rocky peak gazing at the deep solitudes beneath his feet. A cloud approaches, pauses and settles on a projecting ledge. It parts. Mephisto approaches Faust. The two engage in a conversation on

F. W. Murnau's *Faust* (1926).

the origin of mountains, an echo of contemporary controversies between Neptunists and Plutonists in which Goethe had developed a personal interest, especially during his Alpine travels.[17] Yet the conversation soon dies out. Mephisto is exhausted; he seems to have run out of temptations:

> You've looked down, from immeasurable heights,
> On the riches of the world, and its splendid sights.
> Yet, hard as you may be to fire,
> Didn't you feel some deep desire?[18]

Faust yawns at him. They are going nowhere. Gradually, however, Faust begins to change. 'Why should men let things go on being the way they have always been? Isn't it about time for mankind to assert itself against nature's tyrannical arrogance, to confront natural forces in the name of the free spirit that protects all rights?'[19]

Faust reinvents himself once again. He becomes a developer. His appetite is now no longer for theories and visionary dreams, but for concrete operational plans for transforming the earth: 'The boundless nature, how do I make you my own?' As this new vision unfolds, Faust comes back to life. 'Quick, through my mind, leapt plan after plan; / Let rich enjoyment be mine for evermore!'[20] A third peak arises. This time it is a hill generated not by nature, but by human labour. From its top Faust commands his new world in its entirety; a world he has eventually brought into being through massive projects of land reclamation, through vast irrigation networks, through canals, dams and urban planning:

> My gaze revealing, under the sun,
> A view of everything I've done,
> Overseeing, as the eye falls on it,
> A masterpiece of the human spirit,
> Forging with intelligence,
> A wider human residence.[21]

Hazy distant horizons stir Faust's unbound appetite for power and suggest infinite possibilities for further planning and development. For Faust, as for Humboldt (and before him, Saussure), the world has become a vast spectacle constructed for and around the scientist. Yet for Faust, the scientist-developer, ordering nature from the mountaintop is no longer a mere mental process, but one that has tangible impacts on the terrestrial surface and society. The planner's view from above physically transforms the earth; it changes wilderness into ordered, inhabitable land.

The story of Faust betrays Goethe's own fascination with mountains, with a world that appeared to the poet as 'a challenge to the human and its demonic negation'.[22] More significantly, it foreshadows the drama of modernity. While Humboldt's vision was bound to Faust's initial incarnation – the romantic scientist attempting to grasp the hidden forces that animate the cosmos and to find order in nature – Faust moved on. He became himself the active principle *ordering* the world. His world was not simply there to be mapped. The map had become the world.

High-altitude observatories

It is not accidental that Faust's metamorphoses occur on high places. Mountain peaks are not simply evocative settings for romantic drama. They are also privileged platforms for looking at the world, as well as remote charismatic sites set apart from society. As such they hold a transformative potential.

In most religious traditions physical isolation has historically represented a precondition for spiritual enlightenment. Prophets, hermits and shamans fled to mountain wildernesses in search of spiritual quietness, for a more authentic contact with the transcendent and with their inner selves. In ancient and medieval China religious practitioners were drawn to

> mountains that were considered storehouses for potent herbs, plants, and minerals – all of which could be employed in magical spells . . . Mountains served as auspicious places where deities manifested themselves and were therefore

ideal sites to undertake the necessary regimens to attain awakening or ascend as a transcendent.[23]

Likewise, ancient natural philosophers who ventured to lofty peaks were invested with an aura of holiness. Sir John Mandeville, for example, in 1356 described the summit of Athos as the exclusive preserve of 'wise men' who observe the sky, holding a wet sponge on their nose to obtain oxygen at that altitude and inscribing on the ground mysterious letters unwashed by rains or winds.[24]

Separation from the world, whether in the enclosure of a sterile lab or on a remote mountain peak, has remained a necessary condition for the practice of modern science. Isolation allows experiments to be replicated under controlled conditions. At the same time, remoteness and altitude provide unique environments for various kinds of research. High places, for example, have been exploited for the study of meteorology, as well as human physiology under extreme conditions and even to simulate human adaptation to extraterrestrial spaces. Established in 1982 on an over 4,000-m-high glacier (the coldest place in Alaska), the Mount Denali laboratory is used by NASA to explore the dynamics of small isolated technical groups.[25]

More notably, for over a century mountains have constituted favourite sites for modern astronomical observation. In the last decades of the nineteenth century the increase of air and light pollution produced by the Industrial Revolution caused astronomers to move their telescopes from cities to elevated locations with clearer, darker skies. Remote mountain peaks enabled astronomers to work above a significant portion of the earth's atmosphere and its diminished clarity. The Lick Observatory, the first permanent mountaintop astronomical observatory, was founded in 1876 on Mount Hamilton in California at 1,283 m. Observatories at higher altitudes soon followed. The first one was built in 1878 on the Pic du Midi de Bigorre in the French Pyrenees at 2,877 m, with its first dome and telescope installed in 1904. Throughout the century new observatories were established at increasingly elevated locations around the globe, including the

John Mandeville, *Wise men on the Summit of Mount Athos*, c. 1410–20, from *The Travels of Sir John Mandeville*.

Illustration of an astronomical party's ascent of Mont Blanc, from Edward Holden, *Mountain Observatories* (1896).

Sphinx Observatory on Jungfraujoch in Switzerland at 3,571 m, the Chacaltaya Astrophysical Observatory built in 1946 in the Bolivian Andes at 5,230 m and the recent University of Tokyo Atacama Observatory on Cerro Chajnantor in Chile at 5,640 m.

At the turn of the nineteenth century observatories' locations contributed much to the legitimacy of astronomers and their findings. The scientific significance of mountain observatories often seemed to increase in direct proportion to their remoteness and, in a way, it compensated for the impossibility of physically reaching the actual sites of research (whether the moon or Mars). European and American popular press portrayed astronomers in a fashion similar to intrepid glacial explorers moving through treacherous vertical landscapes. The observatories, their final destinations, were always left out of the picture as to emphasize the fierceness of the setting.[26]

The technological sublime

Sphinx Observatory on Jungfraujoch in Switzerland, at 3,571 m.

In the following decades, images of mountain wilderness, masculinity and heroism increasingly fused with technology to produce a new aesthetic taste. Rather than obliterating or downplaying scientific equipments and new technologies in order to exalt nature's purity and roughness, the 'technological sublime' elevated the nature-machine ensemble. The observatory became part of the picture.

The fusion between technological power and natural grandeur had already captured the Western imagination in the late nineteenth century, as railways started to penetrate the American West and the first transoceanic telegraph cables were being laid under the Atlantic. Since the 1920s, however, wireless communications and, above all, powered flight further boosted faith in scientific and technological progress. As with Goethe's Faust, modern man was to challenge and eventually tame wilderness – and mountains.

Storm over Mont Blanc (1930), Arnold Fanck's first sound film, best exemplifies the aesthetics of the technological sublime.

Scenes from Arnold Fanck, *Storm over Mont Blanc* (1930).

The protagonist, Hanes, is the guardian of the remote Mont Blanc observatory. His only contact with the world is through a radio and a small aeroplane that occasionally flies over the refuge. All around are the wild sublime sceneries that entranced and killed generations of mountaineers. Hanes's silent routine is split between housekeeping, recording his observations at the weather station, looking through the telescope, establishing radio contact with other scientists in the valley and, in his spare time, observing minerals through a microscope or simply contemplating the majesty of the landscape while smoking his pipe.

As with Fanck's other *Bergfilmen*, the movie is dominated by romantic views of craggy snow-covered rocks and floating clouds, deep crevasses and scary abysses. The silence of untamed nature is occasionally interrupted by the small plane's engine and by the Morse signals from a large ground observatory. Hanes, the hardy muscular scientist, has become inseparable from technology. And it is through science and technology that his salvation shall come. In the midst of a fierce snowstorm his frozen hands are unable to light fire, but manage to send an SOS signal through his transceiver. A small plane takes off from the valley and heroically battles the blizzard to reach Hanes.

Besides the desperate battle against the elements fought by its protagonists and their devices, in the movie is embedded another technological challenge to wilderness: its very filming. Fanck revered technology as much as he revered nature and always carried the most advanced machinery available with him to the filming site.[27] The taking of shots at high altitude could last several days and the filming process became a real struggle, or, in the words of one of his crews, 'a torture' (indeed, the hut on Mont Blanc gained fame as 'Fanck's concentration camp').[28]

As with the protagonists of other *Bergfilmen*, those of *Storm over Mont Blanc* return from the mountain transformed. They have become wiser and spiritually enriched. Even if in the movie the mountain retains its supremacy – Mont Blanc takes the life of the professor responsible for the ground observatory and it nearly takes Hanes's too – wild nature is ultimately tamed through the lens of the slow-motion camera.

Only three years after the production of *Storm over Mont Blanc*, the first aeroplane flew over Everest and photographed it from above. The purpose of the flight, writes the novelist John Buchan in the foreword to the account of the Houston-Everest expedition, 'was austerely scientific'. The idea was to employ air and photographic technology to gain new knowledge of the area around the peak. A series of survey strips based upon photographs allowed the assemblage of 'a map strip some twenty miles long and something under two miles wide, culminating in the summit of Everest'.[29] Various attempts had been made in previous years, but, as Buchan writes, 'Mt Everest remained unconquered by air and would stay invincible . . . until attacked by an engine of superlative power.'[30]

The feat marked the culmination of the history of visual mastery over nature. Science and technology took the observer not only to mountaintops, but *above* the highest point on earth. The enterprise is narrated by its protagonists in a militaristic fashion. No longer does the scientist hero simply challenge wild nature, but he engages a true war against it – a war ruled by tactics and technology:

> Common sense suggested that the subjugation of this giant depended on strategy and tactics no less thoroughly weighed than those employed in any military campaign . . . A defect in the construction of the machines, an error of judgment or a failure of nerve on the part of the pilots, and the result would have been tragedy.[31]

Aesthetic appeal no longer lies in a grand untouched nature juxtaposed to human technologies, but in 'the sublimity of the manmade . . . fast enough, or complex enough to take on the scale and incomprehensibility of nature'.[32] As Buchan observes, the Houston-Everest expedition took place almost in parallel with Mallory's tragic ascent.

Iain Douglas-Hamilton steers his biplane towards Everest, 3 April 1933.

Bridge on the Marmolada, from the commemorative album of the Società Adriatica di Elettricità, 1955.

> The two expeditions were typical of the old and the new, which must always coexist in the world; the one using the last discoveries of science to circumvent time and space; the other, though assisted by science, relying upon the toughness of the human frame and the power of the human limbs.[33]

In Italy the conquest of mountains through technology became part of Fascist rhetoric. This time, however, mountains were mastered not only through the lens of the camera or the power of the engine, but were physically pierced by tunnels, crossed by asphalted roads and bridges, rendered productive through hydroelectric power plants. And productivity equalled modernity. High-tension pylons, dams, artificial basins and cranes became fundamental elements in the new Alpine landscapes celebrated on postcards, posters and commemorative photographs.[34] Conversely, tamed mountain sceneries were brought into the power plant, as in the case of the mosaic decorating the walls of the Achille Gaggia hydroelectric central in Soverzene, near Belluno. Captured from a high oblique angle, here the peaks surrounding the catchment of the Piave river feature no longer as unproductive wilderness or romantic scenery, but as the reservoir of the region; they no longer serve as panoramic platforms, but as utilities controlled from above.

Whether motivated by science or by the simple desire to 'get to the top', or both, Humboldt's South American wanderings as much as the Houston-Everest expedition and those that preceded and followed them were underpinned by a Faustian restless desire to broaden human horizons. Scientific and technological innovation boasted and accelerated this vision and often changed the landscape, as it did in Italy. At the same time, however, it also ended up producing counter-visions.

In the 1920s and '30s, the American conservationist and urban ecologist Benton MacKaye argued for the value of wilderness as an antidote to an over-urbanized society. MacKaye worried about Americans' increasing alienation from nature and was disturbed by the militaristic and imperialistic tropes underpinning arguments about social progress. Foreshadowing the

Detail from a mosaic featuring the Piave river's upper catchment and its surrounding peaks, at the Achille Gaggia power plant in Soverzene, near Belluno, northeast Italy.

Houston-Everest expedition and the bird's-eye view in the Soverzene mosaic, in a 1921 article he offered his readers a view *over* the Appalachian peaks. This time the ecologist did not employ an aeroplane, but an imaginary giant standing high on the skyline along the mountain ridges, his head scraping the floating clouds. 'What would the giant see from there?' asks MacKaye.

> Starting out from Mt Washington, the highest point in the northeast, his horizon takes in one of the original happy hunting grounds of America – the 'Northwoods' . . . Stepping across the Green Mountains and the Berkshires to the Catskills, he gets his first view of the crowded east – a chain of smoky bee-hive cities extending from Boston to Washington and containing a third of the population of the Appalachian drained area.[35]

The giant then moves his gaze across Pennsylvania, where he notes more smoky columns, 'the big plants between Scranton and Pittsburgh that get out the basic stuff of modern industry – iron and coal'. He then pushes through into the wooded wilderness of the southern Appalachians and proceeds along the great divide of upper Ohio, where he notices 'flowing to waste, sometimes in terrifying floods, waters capable of generating untold hydro-electric energy and of bringing navigation to many a lower stream'.[36]

Mountains here continue to serve their function as scientific observatories. This time, however, they are lookouts not only on nature, but on society. What the giant sees from his Appalachian observatory is what Faust envisaged from his artificial hill. The giant is none other than Faust – the developer, the modern man. 'Resting now on the top of Mt Mitchell, the highest point east of the Rockies, he counts up on his big long fingers the opportunities which yet await development along the skyline he has passed.'[37]

For MacKaye the view from above, however, does not simply allow the study of the dynamic forces of nature and the possibilities for exploitation and development, as in the fascist Dolomites, but provides an 'enlightened perspective' on the sustainability of the land and its people. MacKaye urges Americans to take a break from their hectic lives in the industrial centres and recreate themselves in the mountains, for, he notes, with the exception of the prairies of the Central States, the country's remaining wild areas are for the most part in high places: the Sierras, the Cascades and the Rocky Mountains of the west and the Appalachian Mountains of the east.

The Appalachian range is a particularly convenient area, positioned as it is at less than a day's drive from the most populated areas of the country. With proper facilities, argues MacKaye, these areas 'could be made to serve as the breath of a real life for the toilers in the bee-hive cities along the Atlantic seaboard and elsewhere'. The facilities would include leisure camps connected by a trail running across the entire ridge. MacKaye's vision thus concludes with another Faustian image:

> The spare time for one per cent of our population would be equivalent, as above reckoned, to the continuous activity of some 40,000 persons. If these people were on the skyline, and kept their eyes open, they would see the things that the giant could see. Indeed this force of 40,000 would be a giant in itself. It could walk the skyline and develop its various opportunities.[38]

Even though MacKaye inverts the parts and calls mechanized civilization 'wilderness', his vision ultimately remains utterly modern; it is the vision of a rational planner dreaming from a mountaintop 'a new arcadia that would reunite people and technology in nature'.[39] This vision is nevertheless part of a longer story of conservationism, to which the next chapter will turn.

7 Mountains and Heritage

While it is impossible to put a mountain in a showcase, and even to set boundaries around it, in our mind mountains remain well-defined objects: physical objects for us to experience, aesthetic objects for us to consume, as well as fragile objects to preserve and restore – objects conceptually akin to museum artefacts. As the following pages will show, mountain restoration is a modern invention which developed alongside museums and only became possible once mountains had become 'objects' – beautiful objects.

Yet how is a mountain an object different from, or similar to, those displayed in a museum? Why do we regard mountains as beautiful objects? When did we start to do so? And, more significantly, what are the consequences of doing so? How have mountains helped shape modern environmental consciousness?

Mountains as works of art

In 2002 Mount Kerdylion, a craggy cliff in northern Greece, made headlines. Anastasios Papadopoulos, a Greek American sculptor, launched a campaign to carve it into a 73-m-tall likeness of Alexander the Great – 'four times the size of the presidents of Mount Rushmore'. It was intended as a tribute to the memory of the man who 'brought Hellenism [and thus civilization] throughout the known world'.[1] The colossal sculpture was also meant as a way to proclaim, once and for all, Macedonia's Greekness at a time when Greek pride was smarting over the Macedonian question.

John Muir and Theodore Roosevelt in Yosemite, 1903.

The project sparked curiosity as well as indignation. Environmental groups threatened legal action 'to protect the pine-clad province from being turned into a theme park'. Greek archaeologists and conservationists likewise defended the classical ideal of equilibrium and saw in the project a monstrosity, 'the quintessential example of what Greek tradition is not about: big'.[2]

The irony is that according to Vitruvius, Alexander the Great refused a similar project – and not very far from Kerdylion. His architect, Dinocrates, planned to carve Mount Athos into a giant statue of his patron holding a large city (Alexandria) in his left hand and a bowl that would receive the waters of all the streams on the mountain in his right hand. Alexander's refusal was taken by later classical writers (and contemporary environmentalists) as a demonstration of rationality, as opposed to barbarian hubris.

Megalomaniac or ingenuous as it might sound, Papadopoulos' project raises interesting questions about environmental conservation and about the moral and aesthetic value we ascribe to mountains in general. For Papadopoulos' opposers, Kerdylion

Johann Bernhard Fischer von Erlach's Dinocratic Athos, in *Entwurf einer historischen Architectur* (An Outline of Historical Architecture, 1712).

should not have been turned into a human craft. As a valuable artwork of nature, it was to be preserved as such. Yet the mountain had already been turned into a work of craft – by human imagination.

This is the result of a two-century-long process. While the idea of mountains as beautiful geographical features has ancient roots and is not uncommon in non-Western traditions – from Kulun and its jade palace to emerald Mount Qaf and the harmonious mountain chains decorating medieval Islamic maps, or Korean mountains often described as melodies – in Europe mountains started to become appreciated as artistic objects in their own right rather recently. Between the late eighteenth and nineteenth centuries the 'confused heaps of stones' and 'shapeless and ill-figured old rocks' decried by the theologian Burnet acquired more and more definite form. It is at this time that Mont Blanc emerged on the map as a peak from an indistinct mass of glaciers or *montagnes maudites*. By the end of the century mountains had become paradigms of perfection.[3]

Nineteenth-century European scholars and travellers talked about the mountains of Greece as classical sculptures. For example, in his *Lectures on the Geography of Greece*, delivered at Oxford in 1872, Henry Fanshawe Tozer described the mountains of Attica as artistic creations demanding 'the same education which the mind has to go through in order thoroughly to appreciate a Greek statue or temple'.[4] Others called Greek mountains 'objects of the past', 'nowhere colossal in magnitude', but 'moderated', like the 'Hellenic mind' and like Hellenic sculptures and architecture. The French geographer Élisée Reclus compared Mount Athos in northern Greece to a sphinx and its pinnacle to the top of a high obelisk, like those recently installed in the squares of the great European capitals, whereas David Urquhart described it as a mighty 'column supporting a roof of clouds'.[5] Saussure himself compared the Aiguille du Midi in front of Glacier Montanvert (a section of the Mer de Glace) to an obelisk 'whose slopes are smooth like an artwork'.[6]

The most evocative conceptualization of mountains as artworks, however, is found in John Ruskin's writings and drawings.

John Ruskin, sketch of the Alps, watercolour on cream paper.

Ruskin repeatedly travelled to the Alps, staying in the Jungfrau region, Fribourg and Chamonix to study the mountains and sketch them. To the subject of mountain beauty he devoted an entire volume of his *Modern Painters* (1856). Art and Alpine geology, his life passions, continuously intersect and intertwine in the book. Drawing on his field observations, Ruskin engaged continuously in discussions with scientists as well as with critics. The great Alpine geologists Saussure and Agassiz, Ruskin claimed, 'had gone to the Alps as I desired to go myself, only to look at them, and describe them as they were, loving them heartily'.[7]

For Ruskin mountains were the epitome of beauty – 'the beginning and the end of every natural scenery.' They seemed 'to have been created to show us the perfection of beauty' and this is reflected in the solid shapes in his drawings and in those by his favourite artist, J.M.W. Turner. The greatness of Turner, Ruskin argues, is that he manages to convey the poetry of landscape without giving up or distorting truth. Turner's drawings of the

Alps, he claims, 'are in landscape, what the Elgin marbles or the Torso are in sculpture'.[8]

Ruskin repeatedly compares mountains to sculptures and their features to obelisks, spires, 'pavements of mobile marble', 'domes of snow' or even to the roof of an old French house or the ancient stones of a castle. One of his favourite peaks, the Matterhorn, which every year attracted thousands of admirers, is described by Ruskin as a sculpture graciously chiselled from a single block. Its cliffs, he writes, 'are an unaltered monument, seemingly sculptured long ago, the huge walls retaining yet the forms into which they were first engraven, and standing like an Egyptian temple – delicate, fronted, softly coloured'.[9]

Architectural terms like 'pinnacle', 'buttress' and 'cornice' had been applied to mountain parts for nearly five centuries. In the early 1870s Leslie Stephen argued that 'it is scarcely possible to describe the wildest mountain scenery without the use of architectural metaphor.'[10] Such metaphors continued to proliferate in Alpine tourist guidebooks, with the development of professional mountaineering in the early twentieth century. Geoffrey Young, the author of the first technical mountaineering

John Ruskin, mountains as architectural features, sketch from *Modern Painters*.

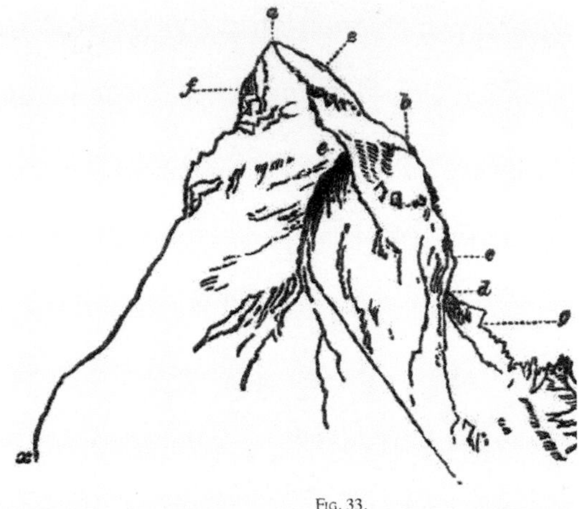

Fig. 33.

Angles with the horizon *x y*.

a f

a e

e b (from point to point)

John Ruskin, sketch of the Matterhorn from *Modern Painters*.

John Ruskin, *Matterhorn from the Moat of the Riffelhorn*, 1849.

guidebook, for example, talks about 'the gothic character of granite and pinnacles'.[11] In Ruskin's writings, however, the metaphor assumes different tones. Mountains embed deep theological and aesthetic meanings, and the two are intimately interconnected. One of Ruskin's most recurrent metaphors is that of 'mountains as Gothic cathedrals':

> [Mountains] are the great cathedrals of the earth, with their gates of rock, pavements of cloud, choirs of stream and stone, altars of snow, and vaults of purple traversed by the continual stars. They seem to have been built for the human race, as at once their schools and cathedrals; full of treasures of illuminated manuscript for the scholar, kindly in simple lessons to the worker, quiet in pale cloisters for the thinker, glorious in holiness for the worshipper . . . Had mankind offered no worship in their mountain churches? Was all that granite sculpture and floral painting done by the angels in vain?[12]

The recurrence of the cathedral metaphor is not accidental. For Ruskin, mountains were the highest and most tangible expression of divine love and their purpose in the divine scheme was threefold: first, mountains served to purify the air; second, they sustained the flowing of rivers; third, and most significantly for Ruskin, they had been created to delight humans, to awaken their poetic and religious consciousness. They were 'as a great and noble architecture; first giving shelter, comfort, and rest; and covered also with mighty sculpture and painted legend'.[13]

As for geologists, for the art critic mountains were magnificent and insistently material objects, but they were also frail and perishable. Like a building, they deteriorated over time. Their histories were ones of endurance and destruction, of eternal decay. Mountains were not the fixed constructions they appeared to be, but ever-changing memorials of how they used to be. Ruskin believed that the geological past was everywhere apparent, if only one knew how to look.[14] In mountains the 'morphological eye' of Ruskin and any other trained observer saw 'shadows of

John Ruskin, *The Northern Arch of the West Entrance of Amiens Cathedral*, 1856, watercolour on paper.

their former selves', ruins of 'splendid forms that were once created'. However, the difference between human architecture and the divine architecture of mountains was that while with the former 'the designer did not calculate upon ruin', with the latter, ruin was part of God's purpose 'and the builder of the temple for ever stands beside His work, appointing the stone that is to fall, and the pillar that is to be abased, and guiding all the seeming wildness of chance and change into ordained splendors and foreseen harmonies'.[15]

Ruskin's conception of landscape and the cosmos was inherently moral and mountains, as objects in the landscape, Ruskin believed, were the most apparent imprints of God's mercy on the face of the earth. The Lord carefully sculpted and chiselled them.

His wisdom was made manifest not only through their perfect forms but in their structure, in 'the materials necessary to make [them] stand'. Their clefts and precipices were shaped by His finger 'as Adam was shaped out of the dust', their ledges were carved upon the earth's surface as the letters were on the Tables of the Law, 'and was it thus left to bear its eternal testimony to His beneficence among these clouds of heaven . . . We should follow the finger of God, as it engraved upon the stone tables of the earth the letters and the law of its everlasting form.'[16]

Mountains did not simply look like cathedrals; the great cathedrals of Europe were also made of mountain stones, as were famous sculptures. The art critic therefore invited his readers to consider the legacy that existed between the mountains of Italy, their different types of minerals and Renaissance artistic genius, as well as 'the kind of admiration with which a southern artist regarded the *stone* he worked in; and the pride which populace or priest took in the possession of precious mountain substance, worked into the pavements of their cathedrals, and the shafts of their tombs'.[17]

Mountains as heritage

Nowadays we still talk about mountains as works of art – perhaps less as sources of precious minerals for sculpture and more and more as valuable crafts to be protected, preserved and displayed to the public in their entireness. Perhaps unconsciously, we continue to ascribe to mountains a sacramental dimension, as Ruskin did. We regard them as 'holy objects' in the most literal sense, as *halig* – something that must be preserved 'whole' or intact; something that, like wilderness, should not be transgressed, violated or altered.

When in 1988 the president of a local commune in the Swiss Canton of Valais proposed to artificially increase the height of Mount Fletschhorn by building a dry stone wall on its top he was fiercely attacked. In the nineteenth century the mountain was estimated to be 4,001 m tall. Later measurements, however, reduced this figure to 3,993 m. The wall would have restored its status as a four-thousander and thus attracted more mountaineers to climb it – some of them had admittedly desisted from doing so precisely because the peak did not fall in that category.

Unlike in the case of Papadopoulos' project, this time the proposal aroused the criticisms not only of scholars and environmentalists but, above all, of common citizens. Several letters of protest appeared in local and national newspapers. A doctor from Germany quoted the biblical passage 'For everyone who exalts himself will be humbled, and he who humbles himself will

Mountaintop removal in West Virginia.

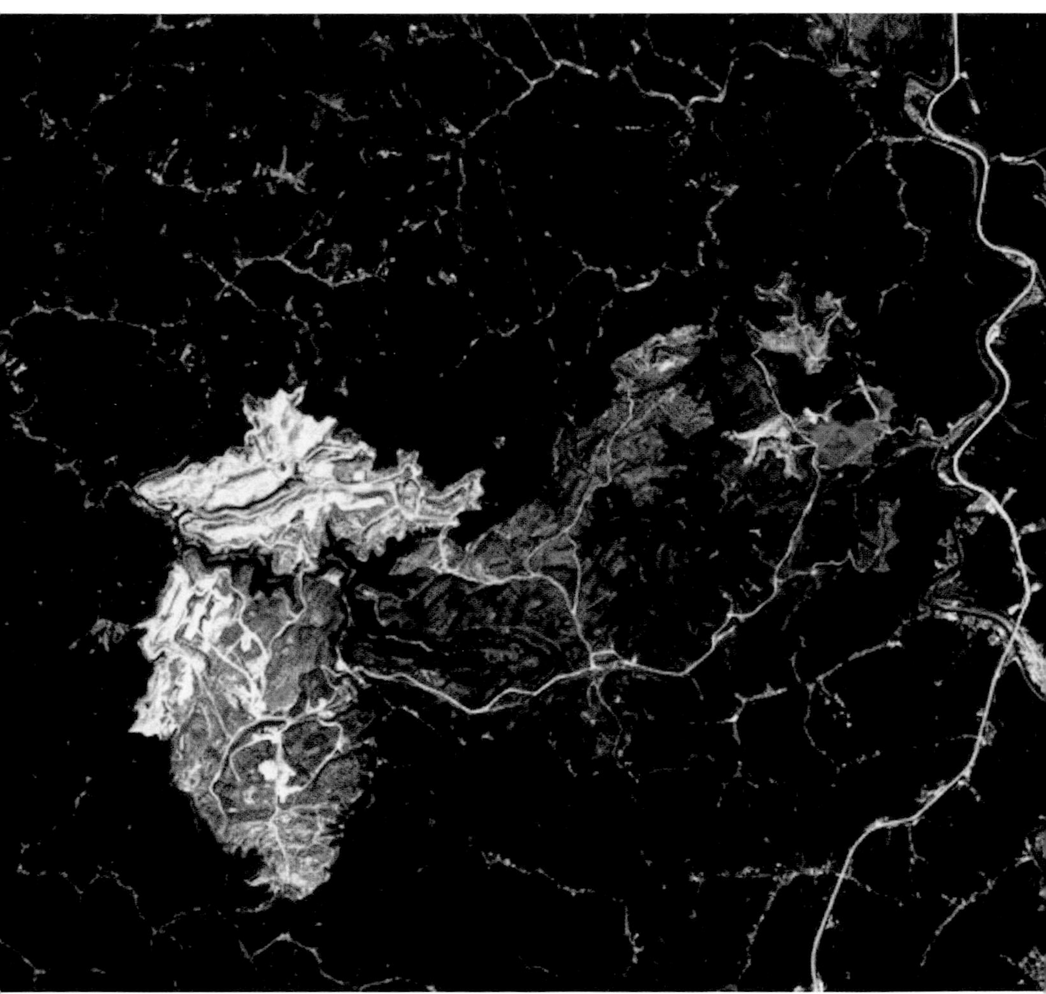

be exalted' (Luke 18:4). Other protesters wrote that they were simply disgusted that 'humans were planning to interfere with God's handiwork'.[18]

On the other side of the Atlantic, similar arguments have been recently made by grassroots associations and other local groups opposing mountaintop removals in the Appalachians. Nowadays mountaintop removal is the predominant method of coal mining in the region. This technique, started in the 1970s, involves blasting the entire summit or summit ridge of a mountain with explosives to expose the coal seams underneath. Each time up to 120 m of mountain is removed, and this excess rock and soil, overloaded with toxic mining byproducts, is often dumped into nearby valleys, which has disastrous impacts on the environment and human health.

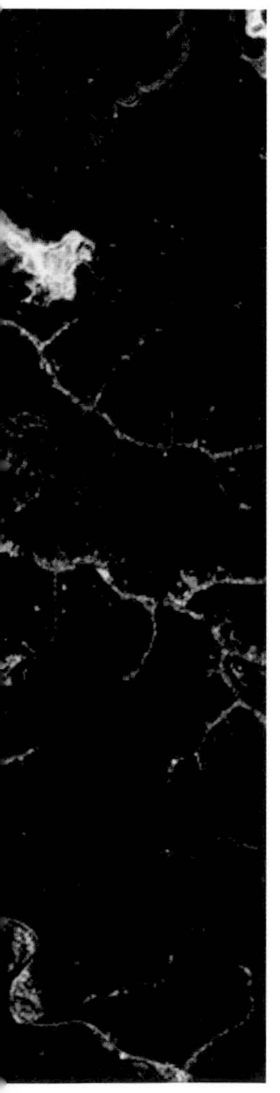

To date, over 470 mountaintops have been blown up in the region. Unlike the Greek and Swiss proposals, here the stake is, of course, much higher. Yet responses are similar. Christians for the Mountains and other organizations that arose in response to mountaintop removal label this practice as a moral and spiritual crisis, even before it is an environmental crisis. The Appalachians are a dramatic symbol of this crisis: 'The order God established, and the task God has given humanity to nurture and protect the creation cannot be abrogated forever. God is not mocked! . . . We humans are wrecking God's creation in central Appalachia and indeed over the whole earth.'[19]

Different as they are, Papadopoulos' project, the Fletschhorn wall and the Appalachian mountaintop removals share some common characteristics. First, mountains are conceived as well-defined objects that can be manipulated and modified. Second, their manipulation is a cause of public outrage. And third, the modification of mountains is perceived as a moral threat, justified by narratives akin to those used by nineteenth-century commentators like Ruskin: narratives of beauty and equilibrium, of holiness and authenticity. But why is the modification of mountains morally unacceptable?

The answer lies in the modern perception of mountains as heritage. Heritage does embue a sacred quality. Its cult, David

Building check dams in the Alps, *c.* 1890.

Lowenthal argues, is a newly popular faith; it is a spiritual vocation, just like nursing or preaching. In the modern West heritage has become 'a self-conscious creed, whose shrines and icons daily multiply and whose praise suffuses public discourse'. Nowadays 'heritage awakens piety the world over . . . [It] answers needs for ritual devotion, especially where other formal faith has become perfunctory or mainly political.'[20]

Environmental restoration and conservation developed side by side with museums and, at least initially, they focused on mountains. Their father, the American diplomat George Perkins Marsh (1801–1881), was inspired by, and in turn inspired, the restoration in the Piedmont Alps. During his time in Italy he observed not only how those mountains had been degraded by their human inhabitants, but how they were being restored through the artificial re-vegetation of their mountainsides and the reconstruction of stream channels – what Cuneo's civil engineers called 'the *restaurazione montana*'.[21] John Muir's environmental struggles similarly took place in the heart of California's Sierra Nevada gold-mining and forest-cutting region. His battles against plans to flood Hetch Hetchy led to its status as a national park in 1906 and as the icon of the struggle for wilderness protection throughout the twentieth century.[22]

Before the nineteenth century natural rather than human agency was deemed the main cause of environmental degeneration. Marsh reversed this assumption and advocated the 'repair' of nature. Restoration, he believed, could undo some of the damage already done. However, environmental conservation and restoration posed (and continues to pose) dilemmas similar to artistic heritage conservation practices. To preserve or to restore? What to preserve? When to restore? To what extent to intervene? How much to reconstruct?

At the very beginning of his influential book *Man and Nature* (1864), Marsh stated that

> man must become a co-worker with nature in the reconstruction of the damaged fabric which the negligence or the wantonness of former lodgers has rendered untenable. He must aid her in re-clothing the mountain slopes with forests and vegetable moulds, thereby restoring the fountains which she provided to water them.[23]

Muir, by contrast, compared wild American mountains and valleys to the most sacred human shrines and advocated their most rigorous preservation. Having climbed a number of mountains in the region, including Cathedral Peak and Mount Dana, he wrote:

> We are now in the mountains and they are in us, kindling enthusiasm, making every nerve quiver, filling every pore and cell of us ... No temple made with hands can compare with Yosemite ... Dam Hetch Hetchy! As well dam for water

> tanks the people's cathedrals and churches, for no holier temple has ever been consecrated by the heart of man.[24]

The exchange of metaphors continued throughout the century. In the 1960s Michelangelo's Sistine Chapel became the man-made equivalent of the Grand Canyon. Two decades thereafter, environmental philosopher Peter Losin argued that 'both the Sistine Chapel and natural systems are better off restored than allowed to decay further'.[25] Others, however, have cautioned that the latter are far more complex than an artwork or a building. Restoring mountains will therefore always be even more difficult than restoring man-made crafts. Restoration is always carried out with the past in mind. Restorers 'see a better past in comparison with a worse present and a hopeful future'. Restoring nature, however, becomes 'even less like its artistic counterpart when one considers the many interpretations of nature and of the ideally restored condition'.[26]

Heritage and its preservation rely on faith rather than on rational proof. Restoration is a matter of subjective judgement rather than an objective truth: 'We elect and exalt our legacy not by weighing its claims to truth, but in feeling that it must be right.'[27] Both artistic and mountain restoration are therefore implemented to different degrees and can take different forms. These often depend on the understanding of the word 'heritage'. The English term places emphasis on the heir (the person who receives), thereby allowing free intervention. The Italian and French words *patrimonio* and *patrimonie*, by contrast, place emphasis on the father (Latin: *pater*), that is, on the giver, thus constraining modifications. The effects can be seen in reconstructions of missing pieces of a monument or historical re-enactments in Britain and North America, as opposed to the non-interventionism in most archaeological sites of continental Europe. They are also seen in tree-planting, as opposed to check dams and soil replacement; in the addition of new materials and the introduction of new plant species; in sustainable land exploitation, as opposed to establishing natural parks; and so on.

Mountains as icons

For Marsh, in order to intervene humans had to think of themselves as separate from nature; they had to turn mountains into objects. For Muir, by contrast, humans were part of nature (and of mountains), yet preserving them intact meant fixing boundaries around protected areas – that is, turning them into islands on the land – another form of objectification. By the early twentieth century, in the Western imagination mountains had become oases of an uncorrupted past in the midst of a changing present, reifications of the deep time of nature, as opposed to the accelerated time of modernity.

The Rockies and the California Sierras became icons of American environmentalism through the paintings of artists such as Thomas Moran and Albert Bierstadt, as well as through the photographs of Eadweard Muybridge and especially of Ansel Adams, the photographer of the Sierra Club. Whether painted on canvas or photographed in monochrome, these images replicate the pictorial conventions of the European picturesque. Their staged symmetrical composition guides the eye towards hidden vanishing points, while at the same time bounding the visual horizon.[28]

Mountains are part of a secret Eden that is progressively disclosed to the viewer and yet at the same time remains safely self-enclosed. Mountains make Yosemite a safe yet precarious refuge from the brutalities of Civil War and the Second World War, as well as from human violence on the environment. Half-concealed by white clouds or exalted by the light of a golden sunset, mountains' solid and yet fragile forms provide an ideal with visual shape. Their dissemination through coffee-table books, exhibitions, posters, brochures and screensavers dramatically contributed to twentieth-century Western (and especially North American) popular perceptions of wilderness as a frail archipelago of parks and protected areas, rather than an unbounded threatening Other.

Since the establishment of Yosemite as a natural reserve, in North America national parks have become sanctuaries to

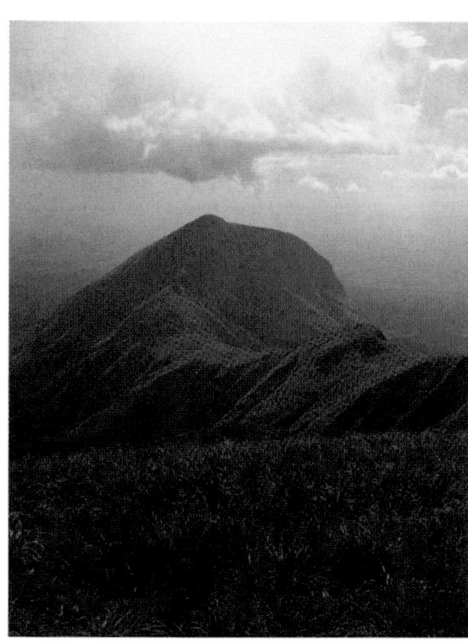

Mount Nimba Strict Nature Reserve.

be preserved for the spiritual benefit of future generations. Nowadays people continue to visit them as pilgrims to 'natural shrines that embody the spirit of the land in its original and pristine condition'. Every year thousands of visitors flock to Yosemite in the same way thousands of visitors flock to the Louvre to see 'the original' *Mona Lisa*.[29]

Similar narratives surround UNESCO mountain World Heritage Sites. A large number of WHS are mountains or are located in mountain regions: from the natural reserve of Mount Nimba (1,752 m), between Guinea and Ivory Coast, to the Dolomites (listed only in 2009), Machu Picchu, the Huangshan range in eastern China and Uluru-Kata Tjuta park in Australia. All these places are conceptualized and treated as uncorrupted islands surrounded by external threats, such as land exploitation, uncontrolled tourism, conflict and climate change. They have become the new icons of environmental and cultural preservation. Mount Nimba, for example, hosts a unique ecosystem and many endemic floral and animal species, but its numerous iron deposits make it a target for exploitation; other sites, by contrast, are endangered by mass tourism. Machu Picchu alone is visited close to 4,000 people per day, causing erosion of the summit and deterioration.[30] In the case of the Dolomites, only the summits are considered World Heritage Sites (the valleys are not).

The most iconic 'insular' WHS, however, is probably Mount Athos. Unlike the Dolomites, the peninsula is defined by natural boundaries for most of its perimeter. It is listed on the basis of both naturalistic criteria (its botanical variety and endemic species) and cultural criteria (its twenty Byzantine monasteries and thriving community of 2,000 monks). Unlike most shrines around the Mediterranean, access to the peninsula is strictly regulated, thus preserving it from mass tourism. Up to 120 male Orthodox Christian pilgrims – women have been banned for the past thousand years – are allowed each day, whereas foreigners

Albert Bierstadt, *Valley of the Yosemite*, 1864, oil on paperboard.

Ansel Adams, *Clearing Winter Storm, Yosemite National Park, California*, 1940.

of other religious affiliations are limited to ten per day. Standard permissions are restricted to three days, during which all visitors receive free hospitality from the monasteries. Exclusiveness, the Mount Athos expert Graham Speake notes, is essential to Athos' survival:

> If it were to be compromised, there is no doubt that within a very short space of time the only surviving holy mountain would suffer the same fate as . . . countless other monasteries in Greece and the Middle East that are now either museums of Byzantine art or deserted ruins.[31]

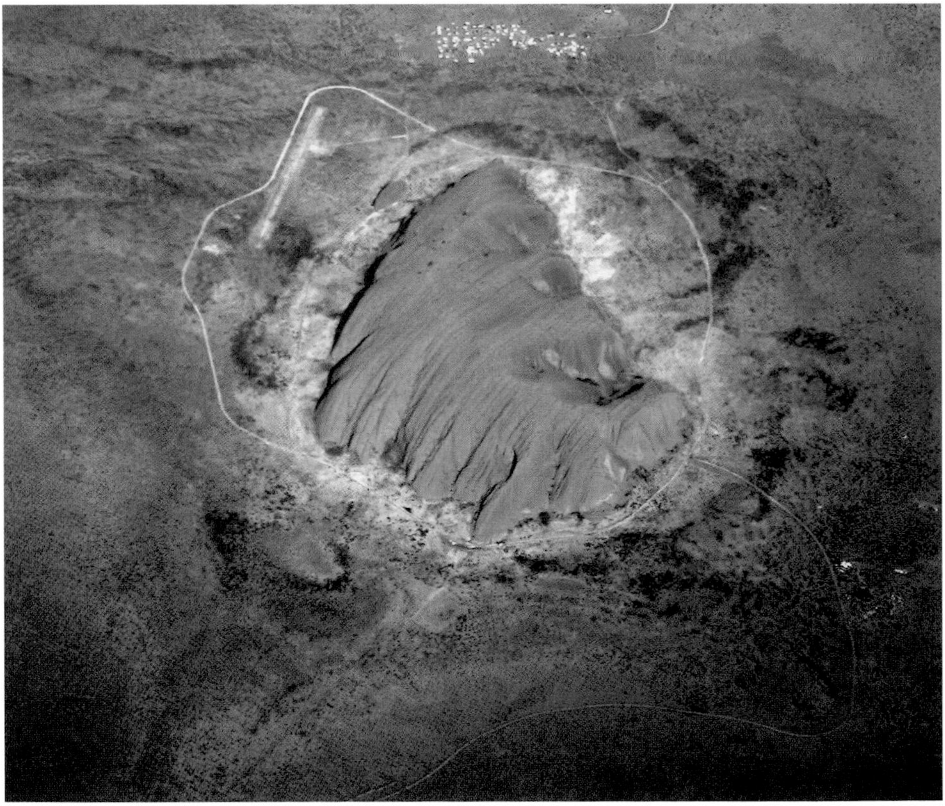

Docheiariou Monastery, Mount Athos, Greece.

Aerial view of Uluru-Kata Tjuta National Park, Australia.

If Athos has become an icon of the preservation of nature, history and spirituality, other mountains have become icons of climate change. With climate change many mountain species are moving upslope, as shifting treelines show in ranges in Europe, Asia, North America and Australia.[32] More notably, contracting glaciers have turned into icons of the fragility of our planet, symbols of environmental catastrophe – and objects under threat. The snow of Kilimanjaro (4,600 m) is disappearing; the glacier of Mount St Elias in Canada has withdrawn 80 km over the past 200 years.[33] Both mountains are inscribed in the UNESCO list of endangered sites, just like the citadel of Bam in

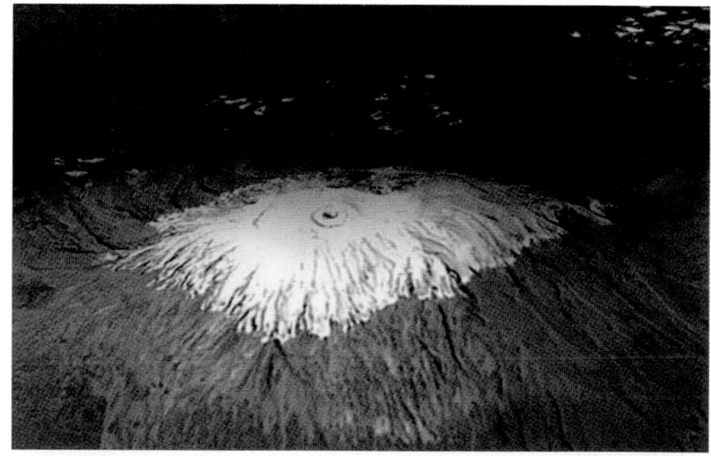

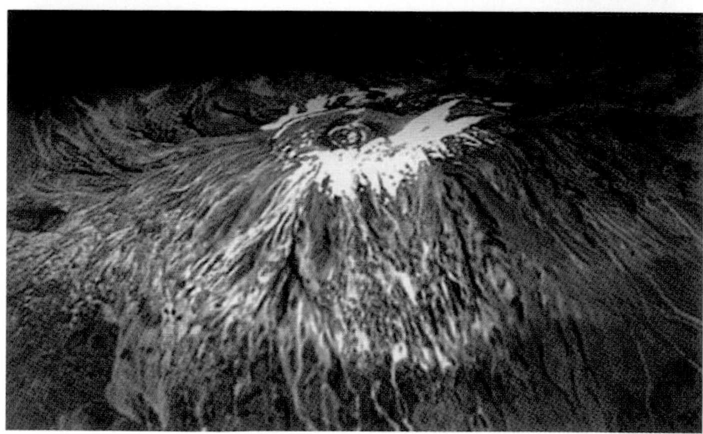

Kilimanjaro's shrinking snowcap in 1993 and 2000.

Iran, the ancient buildings of Zabid in Yemen and other historical monuments under threat.

Mountains on display

Today mountains are no longer simply conceptualized as objects (or islands on the land); they have become commodities. This has parallels in museological practice. In the museum today we search for experiences, rather than for the mere appreciation of artworks. Dioramas, interactive screens and innovative display techniques are making museums increasingly user-friendly. Museums are no longer the preserve of experts, but also of families and weekenders. Likewise, mountains are no longer the exclusive domain of expert mountaineers or naturalists, but of all sorts of excursionists – they have also become user-friendly.

Over the past two centuries mountains have been increasingly perceived as boxes to be ticked, as objects to be classified and collected. In late nineteenth-century Scotland this perception became a true phenomenon known as 'Munroism', the pursuit of the physical ascent of the 283 Scottish 'separate' peaks over 914 m (3,000 ft) listed in Hugh Munro's *Tables* (1891). The accumulation of 'ticked off' peaks enabled (Scottish) mountaineers to establish a close engagement with his or her land, in line with the role of Alpine clubs and nationalist discourses in other parts of Europe. Here, however, the 'scientific' conceptualization of mountains as individual collectible objects was a precondition to such engagement.[34]

Nowadays Dolomite peaks are evocatively framed through the architectural structures of Reinhold Messner's five Mountain Museums. These include a ribbon of glass reflecting snow-clad Ortler peak, carefully staged openings on the walls of ancient castles, and the spectacular panoramic deck on the top of Monte Rite. The museums are not so much concerned with the (contested) local history of those mountains and of the region, as with mountains as a global heritage, regardless of whether they are located in the Dolomites, in the Andes or in the Himalaya. The global peaks ascended by Messner each brand a different

The Firmian Messner Mountain Museum in Sigmundskron Castle near Bolzano, South Tyrol.

product of a cosmetic line available in the museums' stores – perhaps the ultimate way of 'collecting mountains'.[35]

Mountains (like museums) have become objects for mass consumption. The sublime has become a commodity; risk itself has become a commodity. In 2011 a glass skywalk was built on the side of Tianmen Mountain in southeast China. Part of one of the most visited parks of the country, the mountain already boasted the longest mountain cableway in the world, with a total length of 7,455 m. The new glass skywalk, less than a metre wide, is located 1,430 m above sea level. It is built on a precarious sheer rockface and allows crystal-clear views of the terrifying abyss underfoot.[36]

Similar projects have been implemented in the Grand Canyon and Norway. In 1994 the Norwegian Public Roads administration started a project for the development of a network of official scenic routes throughout the country. More recently artistic installations have been set on scenic locations along these routes to facilitate the consumption of sublime mountain views. This 'curatorial' project is part of a longer Norwegian

The Messner Mountain Museum in the Clouds on Monte Rite in the Dolomites.

The Corones Messner Mountain Museum on the Kronplatz mountain, Dolomites.

simonecalo.com

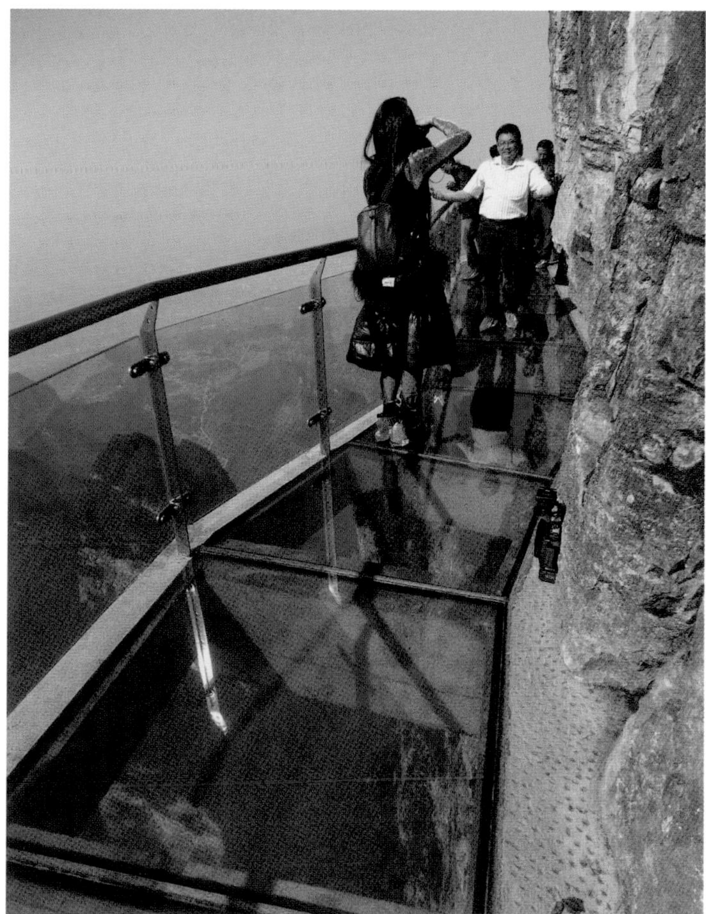

Glass skywalk on Tianmen Mountain in southeast China.

tradition of scenic consumption from dramatic panoramic spots. The new installations largely play with nineteenth-century optical devices and panoramas. Some of them consist of special platforms marking the spot where the visitor is supposed to stand and watch a nature 'objectified and set up for view'. At the same time, they enable the viewer to safely consume the sublime and experience extreme sensations such as vertigo while in safe and comfortable locations.[37]

Saunders Architecture and Tommie Wilhelmsen's *Stegastein*, for example, consists of a large wooden walking board running 33 m off the road into the air. While facing magnificent peaks,

Saunder and Wilhelmsen's *Stegastein* lookout, Aurland, Norway.

it nevertheless directs attention 'more towards the vertiginous sensation of being there than to the expanded views ahead'. As Janike Kampevold Larsen observes, while nineteenth-century panoramas were engineered to give viewers the feeling of being in the middle of the landscape, *Stegastein* strives to create 'the feeling of being *on a platform* in the middle of the landscape, making the viewer aware that her relation to nature is staged'. Installations like the *Stegastein* exaggerate the ravine and place the beholder in 'impossible situations where the architectural structure facilitates and intensifies the experience of drop, height, danger'. They create 'extreme conditions for viewing, while displaying nature'.[38]

By contrast, other installations consisting of glass plates and mirrors enable the viewer to 'enframe wilderness' and turn mountains into literal objects, in the tradition of Claude glasses, the small tinted mirrors used by eighteenth-century artists and travellers for landscape painting. For example, Carl-Viggo Hølmebakk's telescopic viewing device at Nedre Oscarshaug, set on the road crossing the Sognefjell mountain area (the highest pass road in Norway), helps the beholder isolate specific peaks for view while indicating their names on the horizontal plane supporting the glass sheets. Beholders are therefore enabled to know which peaks they are viewing. Mountains are thus not only framed and objectified, but labelled; they are turned into objects for display.[39]

Like smuggled artworks, nowadays mountains have entered global markets as commodities – in the literal sense. The most extreme example comes from Austria. In 2011 two summits in the Carnic

Carl-Viggo Hølmebakk, *Sight Apparatus*, Nedre Oscarshaug, Norway.

View through Hølmebakk's device at Nedre Oscarshaug.

Alps were put on sale to help the government face financial crisis. Gran Kinigat (2,690 m) was listed for 92,000 euros and Rosskopf (2,600 m) for 29,000 euros. The mountains were advertised as 'the most beautiful views in the Carnic Alps, and popular destinations for Alpinists and excursionists'. Twenty customers showed interest, but widespread protest from the general public stopped the plan and the Austrian government eventually suspended the sale. Austrians fiercely opposed the privatization of their mountains. 'This is something that touches us emotionally', said a local mayor, 'because the mountains belong to us like the village church and the forests.'[40]

Even if nothing came of it, the story raises interesting questions: where would boundaries be set? Who would have access to the peaks? Who would be excluded from them? Why have Greek islands been sold and Austrian mountains have not? Is it really possible to set a boundary around a mountain and fix a price to it? What would Ruskin have said?

Epilogue

There are different ways to approach a mountain: from ground level, with fear and reverence; from its top, as a panoramic platform from which to master the surrounding landscape; from above its top, that is, conceptually, as a discrete object; from its slopes, as a pulsating material presence under one's hands and feet; from within its belly, as a dark space of intimate secrecy. Each approach implies a different level of attachment and detachment, of engagement and abstraction, of physical participation and distanced contemplation. While most of these approaches have coexisted throughout human history, some have dominated specific times and cultures.

The conceptualization of mountains as measurable scientific objects and artistic crafts to be preserved is predominantly a product of Western modernity, as is the perspectival view from their top. Yet there is also a counter-history that has been paralleling this process of visual and conceptual mastering: a history of bewilderment, of separation from the ordinary; a history in which time seems to stop. This is the history of Mann's *Magic Mountain* and Shepherd's *Living Mountain*, but it is also the history of the many holy and eerie mountains that have long inhabited our world and continue to haunt our imagination.

Unlike Ruskin, Henry David Thoreau did not see mountains as 'objects' but rather as 'unfinished places'. On Mount Katahdin, a mere 1,606-m-high peak in Maine, the author encountered an 'otherworldly' landscape somehow akin to the Alpine scenery that so much shocked Burnet, a rugged inhospitable landscape where humans did not belong. Rather than a carefully chiselled sculpture, the mountain appeared to him as

> A vast aggregation of loose rocks, as if some time it had rained rocks, and they lay as they fell on the mountain sides, nowhere fairly at rest . . . an undone extremity of the globe . . . This was the Earth of which we have heard, made out of Chaos and Old Night . . . It was the fresh and natural surface of the planet Earth, as it was made forever and ever.[1]

As Ruskin and Thoreau were writing, increasing hordes of tourists were starting to invade the Alps. Mountaineering was turning into spectacle, as clusters of visitors glued their eyes to telescopes set up by the hotels at Chamonix, Zermatt and other climbing centres for the precise purpose of consuming risk from a distance. As the scholar of Victorian literature Ann Colley notes, 'From the moment mountaineers started their ascent to their return they were under surveillance.'[2] Travel accounts, performances (such as those enacted by Albert Smith

Starry night on Mount Everest.

at Piccadilly), panoramas and, since the 1920s, Alpine film all acted as analogues of tourists' telescopes in that they brought perilous mountain ascents and their extreme sceneries into the comfort of urban and even domestic spaces.

The first ascent of Everest and the Cold War added further impulse to both actual and armchair mountaineering. As Geoffrey Young wrote in the mid-1950s, 'The modern lay public is now ready to read mountain adventures among its other sensational readings. It still demands excitement all the time.' However, he continues:

> The cut rope is no longer essential, and the blonde heroine has less appeal, now that she has to climb in nailed boots and slacks. It wants records, above all. Records in height, records in endurance, hair-breadth escapes on record rock walls, and a seasoning of injuries, blizzards, losses of limbs and hazards of life.[3]

Between 1950 and 1964 the world's fourteen eight-thousanders were all ascended for the first time. In 1959 a group of 'real-life heroines' led by Eileen Healey embarked on the first all-female expedition to Cho Oyu in the Himalaya, the world's sixth highest peak. All those feats were duly followed by first-person accounts – such as Maurice Herzog's *Annapurna* (1952), dictated by the author from his hospital bed because he had lost his fingers during the ascent of the eight-thousander, John Hunt's *The Ascent of Everest* (1953), Edmund Hillary's *High Adventure* (1955), Wilfrid Noyce's *South Col* (1954) – and third-person accounts (Stephen Harper's *Lady Killer Peak*, 1962).

Like famous expedition accounts, other adventure books set in the Himalaya proved huge commercial successes. For example, Heinrich Harrer's *Seven Years in Tibet* (1952), the story of the Austrian mountaineer's escape from a British internment camp in India and his extended journey across Tibet, sold three million copies and was translated into 53 languages. The book was then turned into a documentary in 1956 and into a movie in 1997. It was, however, not Harrer's Himalayan best-seller, but his

Swiss counterpart, *The White Spider* (1959), that became the great classic of mountain literature, enthralling generations of climbers. 'I became a mountaineer inspired by the most gripping and frightening mountaineering book I have ever read', confesses Joe Simpson, the author of *Touching the Void* (1988), in his preface to an edition of *The White Spider*.[4]

The book recounts the stories of the various attempts to climb the infamous north face of the Eiger, from Max Sedlmeyer and Karl Mehringer's dramatic endeavour in 1935, followed by other tragedies, to the first successful ascent in 1938 (of which Harrer was part), and Kurt Diemberger and Wolfgang Stefan's triumph in 1958. The title of the book comes from a treacherous spider-shaped glacier which every climber has to cross in order to reach the peak. Here technical abilities and nerves are put to the most extreme of all tests – and 'there is no way around it.'[5]

Characteristic of the book are its continuous shifts of perspective. Stories of successes and disasters usually start to unfold through the lens of tourists' telescopes in the nearby hotels; at moments of heightened drama the author zooms in on the protagonists – now stuck on the wall on the verge of collapse, now precariously dangling on a rope in the midst of a blizzard, now standing triumphant on the summit. The eye then moves back to the telescope; and then up again to the mountain. Through the telescope tragedies become grim live spectacles, and this, Harrer argues, to the detriment of the climbers, who have always longed for solitude. Commenting on the 1957 disaster, he sarcastically remarks:

> The only people immune from error [in the climb] were the googlers who took no part in it. They came in their hundreds more, crowding Grindelwald, Alpiglen and the Kleine Scheidegg to capacity for a whole week, besieged the telescopes, paid inflated prices . . . The weather did all it could for this spectacle, enacted on the most savagely wild natural stage in the whole world; it stayed fine for days on end, and when clouds came up and hid that stage intermittently from sight, they merely provided a welcome

> dropscene, which only served to heighten the suspense of the play . . . Many of the watchers knew well enough that death was in charge of the production, but they wouldn't let that upset them. They only enjoyed a stimulating thrill, while at the same time tasting a modicum of self-satisfaction at the thought that they themselves would have never mixed up in such nonsense.[6]

As with telescopes, Harrer's and other mountaineering books published in the 1950s enabled (and continue to enable) readers to experience vicariously 'the terror and exultation of mountaineering from the warm comfort of their armchairs', leaving them 'haunted with a sense of wonder'.[7]

Mountain films provided a further lens on the mountains, yet in ways that were different from early *Bergfilmen.* From submission to inexorable destiny and elemental might, emphasis now shifted to the rush to the top. As with mountaineering books, in Cold War period Alpine movies, action and 'records' took over contemplation – and politics took over poetry. In American films such as *The White Tower* (1950) and *Third Man on the Mountain* (1959), sons and daughters of mountaineers who perished during their ascents put their lives at risk to fulfil their fathers' dreams, and ideological affiliations work in tension with comradeship.

Inaccessible peaks also turned into settings for thrillers, as rock, snow and death intertwined with stories of espionage and intrigue. Disturbingly, in Clint Eastwood's *The Eiger Sanction* (1975), death ended up transcending fiction and embracing real life. During filming, a young body double and photographer was killed during a rock fall.

Mountain interiors likewise became sites of secrecy and stages for science fiction. Inspired by the NORAD (North American Aerospace Defense Command) facility underneath Cheyenne Mountain in Colorado, John Badham's *WarGames* (1983) features a young hacker who unwittingly accesses a military supercomputer in the mountain programmed to predict possible outcomes of nuclear war. Believing it is a computer game, the boy gets the supercomputer to run the simulation of a thermonuclear war with the USSR. This generates a national nuclear missile scare and almost initiates the Third World War. Seeking to bring a message of peace during the arms race, the film concludes that nuclear warfare is 'a strange game' in which 'the only winning move is not to play'.

Nowadays mountains seem to be within closer reach than ever; and yet their charm has by no means diminished. The demand for mountaineering books, films and documentaries seems to be always on the rise. The figure of the journalist-writer blurs into that of the mountaineer and vice versa, as inquiry, autobiography and artistic experimentation mingle with rock and ice. Jon Krakauer's best-selling *Eiger Dreams* (1990) and *Into Thin Air* (1997) are good examples of this creative fusion, as the writer recounts his attempts to climb iconic peaks, such as the Eiger, and his experience of tragedy on Everest (the 1996 expedition of which he was part saw the death of eight people). The boundary between mountaineering, spectacle and sensation, however, was forever severed when in 1999 the BBC sponsored an expedition to discover evidence of whether Mallory and Irvine had ever reached the summit of Everest before perishing in the blizzard. In the course of the expedition Mallory's body was found. The intersecting stories of Mallory and the mountaineer who retrieved his frozen corpse 75 years later have been recently recounted in Anthony Geffen's movie *The Wildest Dream* (2010).

Conversely, world-renowned climbers have gained fame as writers and actors. Reinhold Messner alone has published over sixty books and has been the protagonist of and contributor to various films and documentaries. Messner was the first man to climb Everest without oxygen (in 1978) and solo (in 1980). He was also the first to climb all the world's eight-thousanders. The success of his books, which have been translated into multiple languages, seems to confirm Young's belief that the modern audience is after 'records', rather than sheer romantic drama. And yet Messner's writing goes beyond the technicalities and the pathos of mountaineering. To him climbing is a philosophy of life:

> For me climbing has [always] been something more than a sport. The danger and difficulties were part of the framework, as much as risk and adventure. Climbing a big wall meant bidding on myself, attracted as I was by a mystery and forced to rely only on myself. Climbing means to move in the open space, to be free to dare to do something outside of the rules, to experiment, to achieve a deeper knowledge of nature.[8]

Messner's philosophy has been recently applied to domains other than mountaineering. In his *Moving Mountains: Lessons on Life and Leadership* (2001) skills and lessons from mountain climbing, such as ascetic concentration and ability to risk, are translated into lessons for successful management. The contemporary mountaineer thus becomes not only writer and journalist, but, self-consciously, a philosopher and master of life. 'Where once was the enthusiasm for the lonely alpine enterprise, today there is the philosophical reflection on alpinism; where once this was stimulated by the experience of a great adventure, nowadays is the preoccupation for ecological balance.'[9]

As snowcaps melt, mountains are increasingly appreciated as objects of nostalgia, as fragments of the 'unhandselled globe', or islands of authenticity dissolving into an uncertain future. At the same time, they continue to perform their role as absolute others, as escapes from the everyday, as sites of separation. And the ultimate act of separation from the ordinary perhaps lies in a shudder. The essence of mountaineering, writes Simpson, is

> that strange mixture of fear and excitement . . . Mountaineering is far, far more than sport . . . It is a nonsensical game of life, and it is this absurd pointlessness that makes it so addictive. If death were not ever present, many would not be drawn to it. Death, in a paradoxical way, validates the life-affirming nature of the game played.[10]

When asked why they put their lives at risk, mountaineers often answer that they are 'possessed' by the mountain. The word

View from Cuzco, Peru.

suggests an impulse beyond rational control; it also evokes the longer biblical association of mountains with epiphany, as well as with hubris and temptation. 'I felt my feet freezing, but paid little attention', recalled Herzog of his ascent of Annapurna:

> The highest mountain to be climbed by men lay under our feet! . . . How many had found on these mountains what, to them, was the end that all mountaineers wish for . . . I was consciously grateful to the mountains for being so beautiful to me that day, and awed by their silence as if I had been in church. I was in no pain and had no worry.[11]

Mountains continue to 'possess' not only their climbers, but those who read their accounts or watch their films from the comfort of an armchair, or consume them through vertiginous yet safe artificial platforms, such as Saunders and Wilhelmsen's *Stegastein* or Tianmen's glass skywalk. Mountains have also come to inspire a new 'aesthetics of difficulty' incarnated by extreme sports, including mountain biking, extreme skiing and wingsuiting, as well as free and solo climbing.[12] These practices free the mountaineer from ropes, as well as from the collective nature of traditional mountain climbing.

Not only do the new sports affirm the definitive autonomy of the modern subject, but they are enacted and mediated through a new type of technology – no longer ropes and telescopes, but social media, headcams and wingsuits. 'Video sharing sites are crammed with mountain material', observes the theatre scholar Jonathan Pitches. In recent years headcams have been used

> to document an almost endless list of signature climbing routes . . . Much of this material is evidence of what has been termed our 'lifecaching' tendency in this century: collecting, storing and displaying one's entire life, for private use, or for friends, family, even the entire world to peruse.[13]

Yet there is something even more disturbing, narcissistic and 'extreme' about these activities. Though they can be highly technical and scientific, extreme mountain sports nevertheless have a high death rate. In this sense, the mountain has become more than an object to be mastered and the summit has become more than 'a secular symbol of effort and reward'.[14] Nowadays mountains linger between commodification and their persisting power to inspire awe, as well as what Edgar Allan Poe called the 'imp of the perverse', the fear of heights motivated by the fear of one's own perverse desire to jump – perhaps the ultimate expression of modernity's tension between the prosaic and the sublime, the mappable and the infinite, the holy and the diabolic, life and death.

THE WORLD'S HIGHEST MOUNTAINS

Mountain	Height (m)	Range	Prominence (m)	First ascent	Ascents (failed attempts) before 2004
Mount Everest	8,848	Mahalangur Himalaya	8,848	1953	>145 (121)
K2	8,611	Baltoro Karakoram	4,017	1954	45 (44)
Kangchenjunga	8,586	Kangchenjunga Himalaya	3,922	1955	38 (24)
Lhotse	8,516	Mahalangur Himalaya	610	1956	26 (26)
Makalu	8,485	Mahalangur Himalaya	2,386	1955	45 (52)
Cho Oyu	8,188	Mahalangur Himalaya	2,340	1954	79 (28)
Dhaulagiri I	8,167	Dhaulagiri Himalaya	3,357	1960	51 (39)
Manaslu	8,163	Manaslu Himalaya	3,092	1956	49 (45)
Nanga Parbat	8,126	Nanga Parbat Himalaya	4,608	1953	52 (67)
Annapurna I	8,091	Annapurna Himalaya	2,984	1950	36 (47)
Gasherbrum I	8,080	Baltoro Karakoram	2,155	1958	31 (16)
Broad Peak	8,051	Baltoro Karakoram	1,701	1957	39 (19)
Gasherbrum II	8,034	Baltoro Karakoram	1,523	1956	54 (12)
Shishapangma	8,027	Jugal Himalaya	2,897	1964	43 (19)

REFERENCES

Preface

1 Mircea Eliade, *The Sacred and the Profane: The Nature of Religion* (New York, 1959).
2 Edwin Bernbaum, *Sacred Mountains of the World* (Berkeley, CA, 1996), p. xv.

1 Mountain Matters

1 The sixteenth-century word 'paramount' in turn derives from the Old French phrase *par amont* ('by above', hence 'superior'). See www.dictionarycentral.com, accessed 15 September 2015.
2 Louise van Swaaij, Jean Klare and David Winner, *The Atlas of Experience* (London, 2000).
3 John Ruskin, *Modern Painters* (London, 1856), pt v, p. 222.
4 Vincent Scully, *The Earth, the Temple, and the Gods: Greek Sacred Architecture* (New Haven, CT, 1962).
5 Bernard Debarbieux and Gilles Rudaz, *Les Faiseurs de montagne* (Paris, 2010); Bernard Debarbieux, 'The Mountain in the City: Social Uses and Transformations of a Natural Landform in Urban Space', *Cultural Geographies*, v (1998), p. 399.
6 Barry Smith and David Mark, 'Do Mountains Exist? Towards an Ontology of Landforms', *Environment and Planning B: Planning and Design*, XXX (2003), pp. 411–27.
7 Jules Blache, *L'homme et la montagne* (Paris, 1933), p. 7.
8 Gino De Vecchis, *Un futuro possibile per la montagna italiana* (Rome, 2004), p. 2.
9 Lowell Thomas, *Lowell Thomas' Book of the High Mountains* (New York, 1969), p. 22.
10 Debarbieux, 'The Mountain in the City', pp. 398–9.
11 Ibid., p. 399.

12 Lisa Sousa, Stafford Poole and James Lockhart, trans. and eds, *The Story of Guadalupe: Luis Laso de la Vega's 'Huei tlamahuiçoltica' of 1649* (Los Angeles, CA, 1998).
13 Jong-Heon Jin, 'Paektudaegan: Science and Colonialism, Memory and Mapping in Korean High Places', in *High Places: Cultural Geographies of Mountains, Ice and Science*, ed. Denis Cosgrove and Veronica della Dora (London, 2009), pp. 196–215.
14 Diderot quoted in Debarbieux and Rudaz, *Les Faiseurs de montagne*, p. 18.
15 Quoted in Avril Maddrell, 'Discourses of Race and Gender and the Comparative Method in Geography School Texts, 1830–1918', *Environment and Planning D: Society and Space*, XVI (1998), p. 84.
16 Michael Bishop and John Shroder, *Geographic Information Science and Mountain Geomorphology* (Berlin and New York, 2004), p. 102.
17 Council of Europe, *Gazette Congress of Local and Regional Authorities of Europe* (May 2000), article 2, 'Definition of Mountain Regions and Territorial Scope', p. 6.

2 Mountains, the Holy and the Diabolic

1 Thomas Burnet, *The Sacred Theory of the Earth: Containing an Account of the Original Creation of the Earth and all the General Changes which It Hath Already Undergone, or is to Undergo till the Consummation of All Things* [1684] (London, 1719), p. 158.
2 Edwin Bernbaum, *Sacred Mountains of the World* (Berkeley, CA, 1996), p. xvi.
3 Edwin Bernbaum, 'Sacred Mountains: Themes and Teachings', *Mountain Research and Development*, XXVI (2006), pp. 304–9.
4 Jon Levenson, *Sinai and Zion: An Entry into the Jewish Bible* (Minneapolis, MN, 1985), p. 116.
5 Bernbaum, *Sacred Mountains of the World*, p. 40.
6 Cited in Edward Casey, *Representing Place: Landscape and Maps* (Minneapolis, MN, 2002), p. 106.
7 Cited in James Robson, *Power of Place: The Religious Landscape of the Southern Sacred Peak (Nanyue) in Medieval China* (Cambridge, MA, 2009), p. 22.
8 Hong-Key Yoon, 'The Role of Pungsu (Geomancy) in Korean Culture', available at https://researchspace.auckland.ac.nz, accessed 15 September 2015; Young Jung, 'Hard Cash for Grave Sales', *Radio Free Asia*, available at www.rfa.org, 5 October 2012.
9 Louise Lamphere, 'Symbolic Elements in Navajo Ritual', *Southwestern Journal of Anthropology*, XXV (1969), pp. 279–305; Malcolm Lewis, 'Maps, Mapmaking and Map Use by Native

North Americans', in *The History of Cartography*, ed. David Woodward and Malcolm Lewis (Chicago, IL, 1998), vol. II, book 3, pp. 51–182.

10 Peregrine Horden and Nicholas Purcell, *The Corrupting Sea* (Oxford, 2001); Alan Peatfield, 'The Topography of Minoan Peak Sanctuaries', *Annual of the British School at Athens*, LXXVIII (1983), pp. 273–9.

11 Vincent Scully, *The Earth, the Temple, and the Gods: Greek Sacred Architecture* (New Haven, CT, 1962), p. 9.

12 Ibid.

13 Gerald Maclean Edwards, *A Greek-English Lexicon* (Cambridge, 1914).

14 Ellen Churchill Semple, *The Geography of the Mediterranean Region: Its Relation to Ancient History* (London, 1932), p. 616.

15 Ibid., pp. 521–2.

16 Mircea Eliade, *The Sacred and the Profane: The Nature of Religion* (London and New York, 1959) p. 53.

17 James Hamilton, *Volcano* (London, 2012), p. 26.

18 Bernbaum, *Sacred Mountains of the World,* pp. 133–5.

19 'Cosmological Image with Mount Meru', catalogue text available at www.metmuseum.org, accessed 15 September 2015.

20 Peter Whitfield, *The Image of the World: Twenty Centuries of World Maps* (London, 2010), p. 22.

21 Kazutaka Unno, 'Cartography in Japan', in *History of Cartography*, ed. J. B. Harley and David Woodward (Chicago, IL, 1987), vol. II, part 2, pp. 372–3; Bernbaum, *Sacred Mountains of the World*, p. 72.

22 Bernbaum, *Sacred Mountains of the World*, p. 85.

23 Mircea Eliade, *The Sacred and the Profane* (Prospect Heights, IL, 1998), p. 231.

24 Belden Lane, *The Solace of Fierce Landscapes: Exploring Desert and Mountain Spirituality* (New York and Oxford, 1998), p. 142.

25 Eric Otto Winstedt, *The Christian Topography of Cosmas Indicopleustes* (Cambridge, 1909), p. 6.

26 Cosmas Indicopleustes, *Topographia Christiana* (6th century AD), 4:185. See also Wanda Wolska, *La topographie chrétienne de Cosmas Indicopleustés: theologie et science au VI siècle* (Paris, 1962). For its contextualization within Byzantine cartography see O.A.W. Dilke, 'Cartography in the Byzantine Empire', in *History of Cartography*, ed. Harley and Woodward, vol. I, pp. 261–2.

27 Procopius, cited in Joseph Hobbs, *Sinai* (Austin, TX, 1995), p. 2; Elisaeus, cited in Lane, *The Solace of Fierce Landscapes*, p. 131.

28 See Solrunn Nes, *The Uncreated Light: An Iconographical Study of the Transfiguration in the Eastern Church* (Grand Rapids, MI, 2007), p. 57.

29 Ibid.
30 Lane, *The Solace of Fierce Landscapes*, p. 137.
31 Andreas Andreopoulos, *Metamorphosis: The Transfiguration in Byzantine Theology and Iconography* (Crestwood, NY, 2005).
32 Veronica Salles-Reese, *From Viracocha to the Virgin of Copacabana* (Austin, TX, 1997), pp. 30–31.
33 Eliade, *The Sacred and the Profane*, p. 216.
34 Christopher Woods, 'At the Edge of the World: Cosmological Conceptions of the Eastern Horizon in Mesopotamia', *Journal of Ancient Near Eastern Religions*, IX (2009), pp. 185–6.
35 Robson, *Power of Place*, p. 19.
36 Ibid., p. 22.
37 Simon Schama, *Landscape and Memory* (New York, 1995), p. 408.
38 Justine Digance, 'Pilgrimage at Contested Sites', *Annals of Tourism Research*, XXX (2003), pp. 143–59.
39 Raphael Jehudah Zwi Werblowsky, 'Introduction: Mindscape and Landscape', in *Sacred Space: Shrine, City, Land*, ed. Joshua Prawer et al. (New York, 1998), p. 16.
40 Karen Pinto, 'Passion and Conflict: Medieval Islamic Views of the West', in *Mapping Medieval Geographies: Geographical Encounters in the Latin West and Beyond, 300–1600*, ed. Keith Lilley (Cambridge, 2013), pp. 201–24.
41 Said Nursi, *The Reasonings: A Key to Understanding the Qur'an's Eloquence* (Somerset, NJ, 2008), p. 58.
42 Karen Pinto, *Ways of Seeing Islamic Maps* (Chicago, IL, in press).
43 Ibid.
44 Giorgio Mangani, 'Rupes Nigra: Mercator and Magnetism', in *Gerhard Mercator: Wissenschaft und Wissentransfer*, ed. Ute Schneider and Stefan Brakensiek (Darmstadt, 2013), pp. 116–31.
45 Bernbaum, *Sacred Mountains of the World*, p. 47.
46 Cited in Schama, *Landscape and Memory*, p. 415.
47 Cited in Jean-Paul Roux, *Montagnes sacrées, montagnes mythiques* (Paris, 1999), p. 134.
48 Ephraim the Syrian, *Hymns on Paradise*, trans. Sebastian Brock (Crestwood, NY, 1990), v. i.4.
49 Ibid., v. i.8.
50 St Gregory of Nyssa, *Life of Moses* (4th century AD), i.5–6.
51 Herbert of Clairvaux, quoted in Sigurdur Thorarinsson, *Hekla: A Notorious Volcano* (Reykjavik, 1970), p. 6.
52 Caspar Peucer, quoted in Thorarinsson, *Hekla*, p. 6.
53 Nicole Chareyron, *Pilgrims to Jerusalem in the Middle Ages* (New York, 2006), p. 105.
54 Frank De Hass, *Buried Cities Recovered: Or, Explorations in Bible Lands, Giving the Results of Recent Researches in the Orient,*

and Recovery of Many Places in Sacred and Profane History Long Considered Lost (Richmond, VA, 1882), p. 297.

55 William Woodhouse, *Aetolia: Its Geography, Topography and Antiquities* [1897] (New York, 1973), p. 32.

56 Betsey Robinson, 'On the Rocks: Greek Mountains and Sacred Conversations', in *Heaven on Earth: Temples, Ritual, and Cosmic Symbolism in the Ancient World*, ed. Deena Ragavan (Chicago, IL, 2013), pp. 175–200.

57 Roux, *Montagnes sacrées, montagnes mythiques*, p. 8.

58 William Harmless, *Desert Christians: An Introduction to the Literature of Early Monasticism* (Oxford and New York, 2004), p. 62.

59 Philip Schaff and Henry Wace, eds, *A Selected Library of Nicene and Post-Nicene Fathers of the Christian Church* (Grand Rapids, MI, 1953), vol. IV, p. 209.

60 Stewart Aubrey, trans., *The Wanderings of Felix Fabri* (London, 1897), vol. IV, p. 587. The same description is found a century earlier in Gervase of Tilbury, *Otia imperialia*, trans. S. E. Banks and James Binns (Oxford, 2002), p. 215.

61 Athanasius of Alexandria, *Vita Antonii* (4th-century AD), 14.7, ed. J. M. Bartelink (Paris, 1994), p. 174.

62 Alice Mary Talbot, 'Les Saintes montagnes à Byzance', in *Le Sacré et son inscription dans l'espace à Byzance et en Occident*, ed. Michel Kaplan (Paris, 2001), pp. 263–76. Byzantine holy mountains operated as focal centres of pilgrimage, as well as crucial resources to which citizens of the Empire turned for spiritual advice and prayer and sometimes also for refuge from political persecution. Peasants, villagers, aspiring nuns, but also generals and sometimes even monarchs challenged the perils of wilderness to consult with the holy fathers, or simply to receive their blessing. The most famous holy mountains were often placed under the protection of the emperor and their foundations sponsored by generous donations. While physically separated from society, holy mountains were an integral and important part of Byzantine spiritual culture, and this is perhaps their main difference from the mountains of Western Europe.

63 Richard Greenfield, ed. and trans., *The Life of Lazaros of Mt. Galesion: An Eleventh-century Pillar Saint* (Washington, DC, 2000), pp. 133–4.

64 Ibid., p. 129.

65 Giorgio Mangani, *Geopolitica del paesaggio: storie e geografie dell'identità marchigiana* (Ancona, 2012).

66 Cited in Bernbaum, *Sacred Mountains of the World*, p. 123.

67 Daniela Querci, 'L'inferno della toponomastica umbra', *Corriere dell'Umbria*, 22 September 2008.
68 Bernbaum, *Sacred Mountains of the World*, p. 142.
69 Ibid., p. xxiii.
70 Eliade, *The Sacred and the Profane*, p. 30.

3 Mountains, Life and Death

1 Jon Krakauer, *Into the Wild* (New York, 1997).
2 Ibid., pp. 56–7.
3 H. J. Fleure, *Human Geography in Western Europe* (London, 1918).
4 J. B. West, 'Highest Permanent Human Habitation', *High Altitude Medical Biology*, III (2002), pp. 401–7.
5 Maurice Herzog, *Annapurna* (London, 1952).
6 Romola Parish, *Mountain Environments* (New York, 2002), pp. 74–5.
7 Jon Mathieu, 'The Sacralization of Mountains in Europe during the Modern Age', *Mountain Research and Development*, XXVI (2006), pp. 343–9.
8 Marjorie Hope Nicolson, *Mountain Gloom and Mountain Glory: The Development of the Aesthetics of the Infinite* (Ithaca, NY, 1959). See also Remo Bodei, *Paesaggi sublimi: gli uomini davanti alla natura selvaggia* (Milan, 2008).
9 Quoted in Gabriel Helland, *Étude biographique et littéraire sur Chênedollé: auteur du poème Le génie de l'homme* (Mortain, 1857), p. 56.
10 Edward Niles Hooker, ed., *The Critical Works of John Dennis* (Baltimore, MD, 1939), vol. II, p. 380.
11 Friedrich Schiller, *Switzerland, St Gotthard: Song of the Alps*, cited in Joel Tyler Headley, *Mountain Adventures in Various Parts of the World: Selected from the Narratives of Celebrated Travellers* (New York, 1876), p. 69.
12 Quoted in Simon Schama, *Landscape and Memory* (New York, 1995), p. 450.
13 Ibid., p. 474.
14 Headley, *Mountain Adventures*, p. 2.
15 Percy Bysshe Shelley, 'Mont Blanc: Lines Written in the Vale of Chamouni', in *The Poetical Works of Coleridge, Shelley and Keats*, ed. Cyrus Redding, A. Galignani and W. Galignani (Washington, DC, 2002), p. 218.
16 William Wordsworth, 'Michael', in *Lyrical Ballads* (1798).
17 Chris Smith, 'Wordsworth and the Mountains', *Alpine Journal* (2005), pp. 238–42.
18 Lowell Thomas, *The Book of High Mountains* (New York, 1969), p. 445.

19 Robert Macfarlane, *Mountains of the Mind: Adventures in Reaching the Summit* (New York, 2004), p. 92.
20 Ibid., p. 88.
21 Headley, *Mountain Adventures,* p. 5.
22 Thomas, *The Book of High Mountains*, p. 447; Edwin Bernbaum, *Sacred Mountains of the World* (Berkeley, CA, 1996), p. 123.
23 Quoted in Headley, *Mountain Adventures*, p. 129.
24 Elizabeth Parker, '"In Memoriam": Obituary of Edward Whymper', *Canadian Alpine Journal* (1912), p. 126.
25 Thomas, *The Book of High Mountains*, p. 450.
26 Macfarlane, *Mountains of the Mind*, p. 225.
27 Quoted in Colonel Norton, 'The Mount Everest Dispatches', *Geographical Journal*, LXIV (1924), p. 164.
28 Eric Shipton, *Men against Everest* (Englewood Cliffs, NJ, 1955), p. 77.
29 Thomas, *The Book of High Mountains*, p. 455.
30 Ibid.
31 Gerry Kearns, 'The Imperial Subject: Geography and Travel in the Work of Mary Kingsley and Halford Mackinder', *Transactions of the Institute of British Geographers*, XII (1997), pp. 240–72.
32 Mary Kingsley, *Travels in West Africa: Congo Francais, Corisco and Cameroons* (London, 1897), p. 550.
33 Halford J. Mackinder, *The Diary of Halford John Mackinder*, in *H. J. Mackinder: The First Ascent of Mount Kenya*, ed. K. M. Barbour (London, 1991), p. 211.
34 Kingsley, *Travels in West Africa*, p. 329.
35 'The Conquest of Ruvvenzori', *The Spectator*, 12 December 1908, p. 21.
36 Frank Mehring, 'Between ngàje Ngài and Kilimanjaro: A Rortian Reading of Hemingway's African Encounters', in *Hemingway and Africa*, ed. Miriam Mandel (Rochester, NY, 2011), p. 216.
37 Bernbaum, *Sacred Mountains*, p. 141.
38 S. Krishna, 'Cartographic Anxiety: Mapping the Body Politic in India', *Alternatives*, XIX (1994), pp. 507–21.
39 Dinesh Kumar, 'Siachen Anniversary: 30 Years of the World's Coldest War', *The Tribune*, 13 April 2014.
40 W.P.S. Sidhu and Pramod Pushkarna, 'The Forgotten War', *India Today,* 13 June 2013.
41 Carl von Clausewitz, *Principles of War*, ed. and trans. Hans W. Gatzke (Harrisburg, PA, 1942), p. 15.
42 Mark Thompson, *The White War: Life and Death on the Italian Front, 1915–1919* (New York, 2009), p. 1.
43 'E il Grappa diventò il Monte Sacro', www.repubblica.it, 28 August 2013.

44 E. Bartlett Kerr, *Flamers over Tokyo: The U.S. Army Air Forces' Incendiary Campaign against Japan, 1944–1945* (New York, 1991), p. 96.
45 Edward Howell, *Escape to Live* (London, 1947), pp. 203–4.
46 Eric Rentschler, 'Mountains and Modernity: Relocating the Bergfilm', *New German Critique*, LI (1990), p. 144.
47 William Bainbridge, 'Heritage in the Clouds: Englishness in the Clouds', PhD thesis, University of Durham, 2014.
48 Rentschler, 'Mountains and Modernity', p. 158.

4 Mountains and Vision

1 Henry Fanshawe Tozer, *A History of Ancient Geography* (Cambridge, 1897), p. 327.
2 Lucian, *The Works of Lucian of Samosata*, trans. H. W. Fowler and F. G. Fowler (Oxford, 1905), vol. I, p. 171.
3 Ibid., pp. 313–14.
4 Tozer, *Ancient Geography*, pp. 324–5.
5 John Wilkinson, trans. and ed., *Egeria's Travels* (Oxford, 2006), p. 121.
6 Ibid., pp. 123–4.
7 Plato, *Phaedo*, 110–15; Clinton Walter Keyes, trans., *Cicero in Twenty-eight Volumes* (Cambridge, MA, 1988), vol. XVI, p. 269; Richard Stoneman, trans., *Pseudo-Callisthenes' Greek Alexander Romance* (London and New York, 1991), p. 123. On the view from above in Stoic tradition, see Denis Cosgrove, *Apollo's Eye: A Cartographic Genealogy of the Earth in the Western Imagination* (Baltimore, MD, 2001) and Denis Cosgrove, 'Globalism and Tolerance in Early Modern Geography', *Annals of the Association of American Geographers,* XCIII (2003), pp. 852–70.
8 Petrarch, 'An Ascent of Mount Ventoux', in J. Robinson, *Petrarch: The First Modern Scholar and Man of Letters* (New York, 1970), p. 308.
9 Ibid., pp. 311–13.
10 Ibid., pp. 313–14.
11 Ibid., pp. 314–16.
12 Ibid., p. 317.
13 Simon Schama, *Landscape and Memory* (New York, 1995), p. 421.
14 Franco Farinelli, *Geografia: Un'introduzione ai modelli del mondo* (Torino, 2003), p. 41.
15 Denis Cosgrove and William Fox, *Photography and Flight* (London, 2010), pp. 16–18; Eric Goldstein, *A Cabinet in the Clouds: J. A. de

Luc, H. B. de Saussure and the Changing Perception of the High Alps, 1760–1810 (Montreal, 2007).
16 Jean Paul Richter, trans., *The Notebooks of Leonardo Da Vinci, Complete* (New York, 1970), vol. I, i. 299, p. 161.
17 Denis Cosgrove and William Fox, *Geography and Vision* (London, 2010), p. 43; Cosgrove, *Apollo's Eye*, p. 128.
18 Schama, *Landscape and Memory*, p. 426.
19 Ibid.
20 John Milton, *Paradise Regained* (London and New York, 1894), book III.261–4 (p. 35).
21 Ibid., III.270–74 (p. 35).
22 Ibid., III.334–6 (p. 36).
23 Peter Hansen, *The Summits of Modern Man: Mountaineering after the Enlightenment* (Cambridge, MA, 2013).
24 Quoted ibid., p. 101.
25 Horace-Bénédict de Saussure, *Le prime ascensioni al Monte Bianco* (Rome, 1981), p. 21. Own translation.
26 Quoted in Hansen, *The Summits of Modern Man*, p. 57.
27 Horace-Bénédict de Saussure, 'Mont Blanc', in Joel Tyler Headley, *Mountain Adventures in Various Parts of the World: Selected from the Narratives of Celebrated Travellers* (New York, 1876), p. 17.
28 Ibid.
29 Hansen, *The Summits of Modern Man*, p. 58. Jean Daniel François Schrader (1844–1924), a French cartographer, geography professor and avid mountaineer who mapped the Pyrenees and was the first to climb the Grand Batchimale (later renamed Pic Schrader), invented the orograph, a device that enabled the mechanical production of these 360° panoramas. Apparently his device never sold, as it was quickly superseded by developments in photography.
30 Alexander von Humboldt, *Cosmos: A Sketch of a Physical Description of the Earth*, trans. E. Otté [1858] (Baltimore, MD, and London, 1997), vol. II, p. 98.
31 Ben Anderson, 'The Construction of an Alpine Landscape: Building, Representing and Affecting the Eastern Alps, *c.* 1885–1914', *Journal of Cultural Geography*, XXIX (2012), p. 163.
32 Yvonne van Eekelen et al., *The Magical Panorama: The Mesdag Panorama, an Experience in Space and in Time* (The Hague, 1996), p. 17.
33 Saussure, *Le prime ascensioni*, p. 25. Rather than avoiding visual contact with the abyss, Saussure suggested, the mountaineer should remain still and gaze at it until he gets used to it. Only then can he proceed. 'However, if he cannot endure the sight of the ravine, he

should give up the feat', p. 131.
34 Anderson, 'The Construction of an Alpine Landscape', pp. 155–83.
35 'Mont Blanc', *Hannibal Journal* (18 March 1852), p. 1.
36 Ann Colley, *Victorians in the Mountains: Sinking the Sublime* (Farnham, 2010), p. 2.
37 Edward Whymper, *Scrambles amongst the Alps in the Years 1860–69* [1900] (New York, 1996), pp. 150–52.
38 Quoted in Edward Casey, *Representing Place: Landscape Painting and Maps* (Minneapolis, MN, 2002), p. 106.
39 Casey, *Representing Place,* pp. 106–9.
40 Jonathan Crary, *Suspensions of Perception* (Cambridge, MA, 2000), p. 288.
41 Nan Shepherd, *The Living Mountain* [1977] (London 2011), p. 59.
42 Ibid., p. 46.
43 Ibid., p. 16.
44 Robert Macfarlane, prologue to Shepherd, *The Living Mountain*, p. xxxi.
45 Shepherd, *The Living Mountain*, p. 98.

5 Mountains and Time

1 'The 10,000 Year Clock', www.longnow.org, accessed 1 December 2013.
2 David Rooney, 'The Clock of the Long Now', www.blog.sciencemuseum.org.uk, accessed 31 July 2013.
3 www.longnow.org/clock.
4 A Taluqdar of Oudh and Srisa Chandra Vasu, *The Matsya puranam* (New York, 1974).
5 Quran, Surah 41:9–10.
6 Thomas Burnet, *The Sacred Theory of the Earth: Containing an Account of the Original Creation of the Earth and all the General Changes which It Hath Already Undergone, or is to Undergo till the Consummation of All Things* [1684] (London, 1719), p. 193.
7 Ibid., p. 194.
8 Eileen Reeves, *Painting the Heavens: Art and Science in the Age of Galileo* (Princeton, NJ, 1997); see also Paola Giacomoni, *Il laboratorio della natura: paesaggio montano e sublime in età moderna* (Milan, 2001), p. 9.
9 Burnet, *The Sacred Theory of the Earth*, pp. 197–9.
10 Ibid., p. 199.
11 Robert Macfarlane, *Mountains of the Mind: Adventures in Reaching the Summit* (New York, 2004), p. 25.

12 Burnet, *The Sacred Theory of the Earth*, p. x.
13 Giacomoni, *Il laboratorio della natura*, pp. 65–75.
14 Quoted in Cathy Gere, 'Inscribing Nature: Archaeological Metaphors and the Formation of the New Sciences', *Public Archaeology*, II (2002), p. 197.
15 Macfarlane, *Mountains of the Mind*, pp. 31–2.
16 James Hutton, *Theory of the Earth* (Edinburgh, 1795), vol. I, p. 15.
17 Ibid., p. 200.
18 John Playfair, *Biographical Account of the Late Dr James Hutton* (Edinburgh, 1805), p. 73.
19 Quoted in Simon Schama, *Landscape and Memory* (New York, 1995), p. 488.
20 Timothy Mitchell, 'Caspar David Friedrich's *Der Watzmann*: German Romantic Landscape Painting and Historical Geology', *Art Bulletin*, LX (1984), p. 459.
21 Rebecca Solnit, *River of Shadows: Eadweard Muybridge and the Technological Wild West* (New York, 2003), p. 13.
22 Stephen Jay Gould, *Time's Arrow, Time's Cycle: Myth and Metaphor in the Discovery of Geological Time* (Cambridge, MA, 1987), p. 151.
23 Charles Lyell, *Principles of Geology, or, the Modern Changes of the Earth and its Inhabitants Considered as Illustrative of Geology* (New York, 1854), p. 405.
24 Ibid., pp. 405–6.
25 Agassiz quoted in Solnit, *River of Shadows*, p. 64.
26 Ibid., p. 13.
27 Yi-Fu Tuan, 'Mountains, Ruins and the Sentiment of Melancholy', *Landscape*, III (1964), p. 30.
28 Quoted ibid.
29 John Gould Fletcher, *Songs of the Rio Grande* (1934), cited ibid., p. 30.
30 Joel Tyler Headley, *Mountain Adventures in Various Parts of the World: Selected from the Narratives of Celebrated Travellers* (New York, 1876), p. 147.
31 Edward Daniel Clarke, *Travels in Various Countries of Europe, Asia and Africa* (London, 1813), p. 71.
32 Ibid., pp. 71–2.
33 Ibid., pp. 73–4.
34 Lord Gordon Byron, 'Childe Harold's Pilgrimage' (London, 1842), I.LX.
35 Robert Eisner, *Travelers to an Antique Land: The History and Literature of Travel to Greece* (Ann Arbor, MI, 1991), p. 71.
36 Quoted in Richard Stoneman, *Land of Lost Gods: The Search for Classical Greece* (London, 1987), p. 144.

37 George Ferguson Bowen, *Mount Athos, Thessaly and Epirus: Diary of a Journey* (London, 1852), p. 47.
38 Henry Fanshawe Tozer, *Lectures on the Geography of Greece* [1882] (Chicago, IL, 1974), p. 43.
39 Robert Shannon Peckham, *National Histories, Natural States: Nationalism and the Politics of Place in Greece* (London, 2001).
40 Angelos Vlachos, *Peri Panagiōtou Soutsou kai tōn poiēseōn: logos apangeltheis en tō Philologikō Syllogō Parnassō tē 22 Martiou 1874* (On Panaghiotes Soutsos and his Poems: Speech delivered to the Philological Association of Parnassus on 22 March 1874) (Athens, 1874), pp. 14–15. Own translation.
41 Quoted in Edwin Bernbaum, *Sacred Mountains of the World* (Berkeley, CA, 1996), p. 231.
42 Yi-Fu Tuan, *Romantic Geography* (Madison, WI, 2013), p. 46.
43 Thomas Mann, *The Magic Mountain*, trans. Lowe-Porter (New York, 1927), p. 541.
44 Ibid., pp. 606–7.
45 Joan Nogué and Joan Vicente, 'Landscape and National Identity in Catalonia', *Political Geography*, XXIII (2004), p. 119.

6 Mountains, Science and Technology

1 Pliny the Elder, *Natural History*, 25.1.
2 Pomponius Mela, *Description of the World*, trans. F. E. Romer (Detroit, MI, 1998), 2.31.
3 Cited in Henry Fanshawe Tozer, *Lectures on the Geography of Greece* [1882] (Chicago, IL, 1974), pp. 323–4.
4 Basil of Caesarea, *Hexamero*, 4th century, 8.8.
5 Cited in Lowell Thomas, *Book of the High Mountains* (New York, 1969), pp. 25–6.
6 Paola Giacomoni, *Il laboratorio della natura: paesaggio montano e sublime in età moderna* (Milan, 2001), pp. 134–7.
7 Richard Grove, *Green Imperialism: Colonial Expansion, Tropical Island Edens and the Origins of Environmentalism, 1600–1860* (Cambridge and New York, 1995), p. 325.
8 Alexander von Humboldt, *Cosmos: A Sketch of a Physical Description of the Earth*, trans. E. Otté [1858] (Baltimore, MD, and London, 1997), vol. I, p. 55.
9 Alexander von Humboldt, *Rélation historique du voyage aux régions équinoxiales du Nouveau Continent* (Paris, 1814), vol. I, p. 31.
10 Cited in Michael Dettelbach, 'Global Physics and Aesthetic Empire: Humboldt's Physical Portrait of the Tropics', in

Visions of Empire: Voyages, Botany and Representations of Nature, ed. David Philip Miller and Peter Hanns Reill (Cambridge, 1996), p. 268.

11 Humboldt, *Cosmos*, vol. I, p. 33.

12 Ibid., p. 79.

13 Ibid., vol. II, p. 97.

14 Edmunds V. Bunkse, 'Humboldt and an Aesthetic Tradition in Geography', *Geographical Review*, LXXI (1981), p. 131.

15 Johann Wolfgang Von Goethe, *Faust*, I.I.438–9, trans. A. S. Kline (2003), www.poetryintranslation.com, accessed 29 January 2016.

16 Ibid., I.II.1075–81.

17 Ibid., IV.I.10040–120.

18 Ibid., IV.I.10130–34.

19 Marshall Berman, *All that is Solid Melts into Air: The Experience of Modernity* (New York, 1988), p. 61.

20 Goethe, Faust, IV.I.1022–8.

21 Ibid., V.I.11245–50.

22 Giacomoni, *Il laboratorio della natura*, p. 15.

23 James Robson, *Power of Place: The Religious Landscape of the Southern Sacred Peak (Nanyue) in Medieval China* (Cambridge, MA, 2009), pp. 20–21.

24 John Mandeville, *The Voiage and Travayle of Sir John Mandeville* (London, 1887), pp. 15–16.

25 Winifred Gallagher, *Power of Place: How our Surroundings Shape Our Thoughts, Emotions and Actions* (New York, 1993), pp. 66–7.

26 Maria Lane, *Geographies of Mars: Seeing and Knowing the Red Planet* (Chicago, IL, 2011).

27 Eric Rentschler, 'Mountains and Modernity: Relocating the Bergfilm', *New German Critique*, LI (1990), p. 146.

28 Peter Hansen, *The Summits of Modern Man: Mountaineering after the Enlightenment* (Cambridge, MA, 2013), p. 236.

29 John Buchan, 'Foreword', in P.F.M Fellowes et al., *First over Everest: The Houston Mount Everest Expedition, 1933* (New York, 1934), p. 13.

30 Ibid., p. 26.

31 Ibid., p. 15.

32 Mari Hvattum, 'The Man Who Loved Views: C. A. Pihl and the Making of the Modern Landscape', in *Routes, Roads and Landscape*, ed. Mari Hvattum et al. (Farnham, 2011), p. 118.

33 Buchan, 'Foreword', to *First over Everest*, p. 15.

34 Gabriele Zanetto, Francesco Vallerani and Stefano Soriani, *Nature, Environment, Landscape: European Attitudes and Discourses in the Modern Period, the Italian Case, 1900–1970* (Padua, 1996), p. 24.

35 Benton MacKaye, 'An Appalachian Trail: A Project in Regional Planning', *Journal of the American Institute of Architects*, IX (October 1921), p. 326.
36 Ibid.
37 Ibid.
38 Ibid., p. 327.
39 Christine Macy and Sarah Bonnemaison, 'The Concept of Flow in Regional Planning: Benton MacKaye's Contribution to the Tennessee Valley Authority', in *Routes, Roads and Landscapes*, ed. Hvattum, p. 147.

7 Mountains and Heritage

1 Anthi Karassava, 'Alexander the Great', *Vintage Radio Cellar* (28 August 2002).
2 Ibid.
3 Edwin Bernbaum, *Sacred Mountains of the World* (Berkeley, CA, 1996), p. 121.
4 Henry Fanshawe Tozer, *Lectures on the Geography of Greece* [1882] (Chicago, IL, 1974), p. 165.
5 David Urquhart, *The Spirit of the East: Illustrated in a Journal of Travel through Roumelia during an Eventful Period* (London, 1838), pp. 178–9.
6 Horace-Bénédict de Saussure, *Le prime ascensioni al Monte Bianco* (Rome, 1981), p. 48.
7 Denis Cosgrove, *Geography and Vision: Seeing, Imagining and Representing the World* (London, 2008), p. 127.
8 John Ruskin, 'Deucalion', in *The Works of Ruskin*, ed. Edward Cook and Alexander Wedderburn, vol. xxvi (London, 1906), p. 102.
9 John Ruskin, *Modern Painters*, vol. v: *Mountain Beauty* (London, 1856), p. 235.
10 Leslie Stephen, *The Playground of Europe* (London, 1871), p. 13.
11 Geoffrey Winthrop Young, *Mountain Craft* (London, 1920), p. 381. See also Anja Karina Nydal, 'Repertoires of Architects and Mountaineers: A Study of Two Professions', PhD thesis, University of Kent, 2013.
12 Ruskin, *Modern Painters*, vol. v, pp. 349–50.
13 Ibid., p. 91.
14 Cosgrove, *Geography and Vision*, pp. 121–34.
15 Ruskin, *Modern Painters*, vol. v, p. 142.

16 Ibid., p. 90.

17 To various extents, argued Ruskin, the very genius of great Italian artists had been 'shaped by mountains' – by their materials and by their aesthetics: 'The sculpture of the Pisans was taken up and carried into various perfection by the Lucchese, Pistojans, Sienese and Florentines. All these are inhabitants of truly mountain cities, Florence being as completely among the hills as Inspruck is, only the hills have softer outlines. Those around Pistoja and Lucca are in a high degree majestic. Giotto was born and bred among these hills. Angelico lived upon their slope. The mountain towns of Perugia and Urbino furnish the only important branches of correlative art' (ibid., p. 358).

18 Jon Mathieu, 'The Sacralization of Mountains in Europe during the Modern Age', *Mountain Research and Development*, XXVI (2006), p. 343.

19 Christians for the Mountains, *The Mountain Vision: The Quarterly Publication of Christians for the Mountains*, I (2013), p. 1.

20 David Lowenthal, *The Heritage Crusade and the Spoils of History* (Cambridge, 1998), pp. 1–2.

21 Marcus Hall, *Earth Repair: A Transatlantic History of Environmental Restoration* (Charlottesville, VA, 2005), p. 55. 'We go much into the Alps, being tempted there by the exceeding proximity of them', Marsh wrote from his villa in Piombino. 'In fact, we are so near that I often amuse myself, as I sit on my balcony . . . knocking the icicles off the caves of that respectable hillock, Monte Rosa, by shying pebbles at 'em. Nay, when it is very clean, I can reach the walls of the Alps with my pipe stem' (ibid., p. 54).

22 Denis Cosgrove, 'Images and Imagination in 20th-century Environmentalism: From the Sierras to the Poles', *Environment and Planning A*, XL (2008), p. 1866.

23 Cited in Hall, *Earth Repair*, p. 7.

24 Cited in Bernbaum, *Sacred Mountains of the World*, p. 252.

25 Cited in Hall, *Earth Repair*, p. 2.

26 Ibid., p. 3.

27 Lowenthal, *The Heritage Crusade*, p. 2.

28 Yosemite had been at the heart of the Sierra's gold-mining and forest-cutting region. 'By 1900 its meadows were heavily grazed by sheepherders and its waters subject to the urban demands of San Francisco. John Muir's battles against ranchers and engineers' plans to flood Hetch Hetchy led to national park status in 1906 and became the reference point for struggles over wilderness protection throughout the twentieth century.' In Moran's and Bierstadt's paintings, mountains are presented as part of a new Eden, as

an idealized refuge for a nation brutalized by the Civil War. In Adams's pictures the same pictorial conventions are preserved and magnified by monochrome (Cosgrove, 'Images and Imagination', p. 1866).

29 Bernbaum, *Sacred Mountains of the World*, p. 252.
30 Alonzo C. Addison, *Mundos en extinción: los lugares más hermosos del planeta en peligro de desaparecer* (Barcelona, 2008), pp. 56, 82.
31 Graham Speake, *Mount Athos: Renewal in Paradise* (New Haven, CT, 2002), p. 266.
32 Jonathan Adams, *Species Richness: Patterns in the Diversity of Life* (Berlin and New York, 2009), p. 297.
33 Addison, *Mundos en extinción*, pp. 238–44.
34 Hayden Lorimer and Kathryn Lund, 'Performing Facts: Finding a Way over Scotland's Mountains', in *Nature Performed*, ed. Bronislaw Szerszynski, Wallace Heim and Claire Waterton (Oxford, 2003), pp. 130–44.
35 William Bainbridge, 'Heritage in the Clouds: Englishness in the Clouds', PhD thesis, University of Durham (2014), p. 458.
36 Stephanie Darrall, 'Don't Look Down! The Terrifying See-through Path Stuck to a Chinese Cliff-face 4,000 ft above a Rocky Ravine', www.dailymail.co.uk, accessed 20 September 2015.
37 Nina Frang Hoyum and Janike Kampevold Larsen, eds, *Views: Norway Seen from the Road, 1733–2020* (Oslo, 2012), pp. 177–94.
38 Janike Kampevold Larsen, 'Curating Views: The Norwegian Tourist Route Project', in *Routes, Roads and Landscape*, ed. Mari Hvattum et al. (Farnham, 2011), pp. 184–6.
39 Ibid., p. 184.
40 'Due cime delle Alpi in vendita: succede in Tirolo', www.mountainblog.it, accessed 15 September 2015.

Epilogue

1 Henri David Thoreau, *The Maine Woods*, ed. Joseph Moldenhauer (Princeton, NJ, 1972), pp. 63–70. See also Nicholas Entrikin, 'The Unhandselled Globe', in *High Places: Cultural Geographies of Mountains, Ice, and Science*, ed. Denis Cosgrove and Veronica della Dora (London, 2009), pp. 216–26.
2 Ann Colley, *Victorians in the Mountains: Sinking the Sublime* (Farnham, 2010) p. 59.
3 Quoted in Heinrich Harrer, *The White Spider* [1959] (London, 2005), p. 14.
4 Joe Simpson, Foreword, in Harrer, *The White Spider*, n.p.
5 Harrer, *The White Spider*, p. 21.

6 Ibid., pp. 208–9.
7 Simpson, foreword to Harrer, *The White Spider*, n.p.
8 Reinhold Messner, *La mia vita al limite: autobiografia di una leggenda dell'alpinismo* (My Life at the Extreme: Autobiography of a Legend in Mountaineering) (Milan, 2006), p. 3.
9 Herbert Henzler, foreword to Reinhold Messner, *Spostare le montagne: come si affrontano le sfide superando i propri limiti* (Moving Mountains: How to Face Challenges Overcoming our Own Limits) (Milan, 2011), p. 2.
10 Simpson, foreword to Harrer, *The White Spider*, n. p.
11 Quoted in Robert Macfarlane, *Mountains of the Mind: Adventures in Reaching the Summit* (New York, 2004), p. 8.
12 Anja Karina Nydal, 'Repertoires of Architects and Mountaineers: A Study of Two Professions', PhD thesis, University of Kent (2013).
13 Jonathan Pitches, 'Deep and Dark Play in the Mountains: Daring Acts and their Retelling', *Moving Mountains: Studies in Place, Society and Cultural Representation*, Conference presentation, University of Edinburgh, History of Art and ESALA, 18–20 June 2014.
14 Macfarlane, *Mountains of the Mind*, p. 142.

SELECT BIBLIOGRAPHY

Bernbaum, Edwin, *Sacred Mountains of the World* (Berkeley, CA, 1996)

Bishop, Michael, and John Shroder, *Geographic Information Science and Mountain Geomorphology* (Berlin and New York, 2004)

Bodei, Remo, *Paesaggi sublimi: gli uomini davanti alla natura selvaggia* (Milan, 2008)

Burnet, Thomas, *The Sacred Theory of the Earth* (London, 1684)

Colley, Ann, *Victorians in the Mountains: Sinking the Sublime* (Farnham, 2010)

Cosgrove, Denis, and Veronica della Dora, eds, *High Places: Cultural Geographies of Mountains, Ice, and Science* (London, 2009)

Debarbieux, Bernard, and Gilles Rudaz, *Les Faiseurs de montagne* (Paris, 2010)

della Dora, Veronica, *Imagining Mount Athos: Visions of a Holy Mountain from Homer to World War II* (Charlottesville, VA, 2011)

Diemberger, Kurt, *Summits and Secrets* (London, 1971)

Eliade, Mircea, *The Sacred and the Profane: The Nature of Religion* (New York, 1959)

Fellowes, Peregrine Forbes Morant, et al., *First over Everest!: The Houston-Mount Everest Expedition, 1933* (New York, 1934)

Giacomoni, Paola, *Il laboratorio della natura: paesaggio montano e sublime in età moderna* (Milan, 2001)

Gould, Stephen Jay, *Time's Arrow, Time's Cycle: Myth and Metaphor in the Discovery of Geological Time* (Cambridge, MA, 1987)

Hansen, Peter, *The Summits of Modern Man: Mountaineering after the Enlightenment* (Cambridge, MA, 2013)

Harrer, Heinrich, *Seven Years in Tibet* (New York, 1954)

—, *The White Spider* [1959] (London, 2005)

Headley, Joel Tyler, *Mountain Adventures in Various Parts of the World: Selected from the Narratives of Celebrated Travellers* (New York, 1876)

Hemingway, Ernest, *The Snows of Kilimanjaro, and Other Stories* (New York, 1927)
Herzog, Maurice, *Annapurna* (London, 1952)
Hillary, Edmund, *High Adventure* (New York, 1955)
Hunt, John, *The Conquest of Everest* (New York, 1953)
Krakauer, Jon, *Eiger Dreams: Ventures among Men and Mountains* (New York, 1997)
—, *Into the Wild* (New York, 1997)
—, *Into Thin Air* (New York, 1997)
Lane, Belden, *The Solace of Fierce Landscapes: Exploring Desert and Mountain Spirituality* (New York and Oxford, 1998)
Lyell, Charles, *The Principles of Geology*, 3 vols (London, 1830–33)
Macfarlane, Robert, *Mountains of the Mind: Adventures in Reaching the Summit* (New York, 2004)
Mann, Thomas, *The Magic Mountain*, trans. H. T. Lowe-Porter (London, 1924)
Messerli, Bruno, and Jack Ives, eds, *Mountains of the World: A Global Priority* (New York, 1997)
Messner, Reinhold, *La mia vita al limite: autobiografia di una leggenda dell'alpinismo* (Milan, 2006)
Nicolson, Marjorie Hope, *Mountain Gloom and Mountain Glory: The Development of the Aesthetics of the Infinite* [1959] (Seattle, WA, 1997)
Parish, Romola, *Mountain Environments* (New York, 2002)
Petrarch, Francesco, 'An Ascent of Mount Ventoux', in J. Robinson, *Petrarch: The First Modern Scholar and Man of Letters* (New York, 1970), pp. 307–20.
Robson, James, *Power of Place: The Religious Landscape of the Southern Sacred Peak (Nanyue) in Medieval China* (Cambridge, MA, 2009)
Roux, Jean-Paul, *Montagnes sacrées, montagnes mythiques* (Paris, 1999)
Saussure, Horace-Bénédict de, *Le prime ascensioni al Monte Bianco* (Rome, 1981)
Schama, Simon, *Landscape and Memory* (New York, 1995)
Shepherd, Nan, *The Living Mountain* [1977] (London, 2011)
Shipton, Eric, *Men against Everest* (Englewood Cliffs, NJ, 1955)
Smith, Chris, 'Wordsworth and the Mountains', *Alpine Journal* (2005), pp. 238–42
Stephen, Leslie, *The Playground of Europe* (London, 1871)
Talbot, Alice Mary, 'Les Saintes montagnes à Byzance', in *Le Sacré et son inscription dans l'espace à Byzance et en Occident*, ed. Michel Kaplan (Paris, 2001), pp. 263–76
Thomas, Lowell, *Book of the High Mountains* (New York, 1969)
Thoreau, Henry David, *The Maine Woods*, ed. Joseph Moldenhauer (Princeton, NJ, 1972)

Tuan, Yi-Fu, *Romantic Geography: In Search of the Sublime Landscape* (Madison, WI, 2013)
Whymper, Edward, *Scrambles amongst the Alps in the Years 1860–69* [1900] (New York, 1996)

ASSOCIATIONS AND WEBSITES

Alpine Club
www.alpine-club.org.uk

Alpine Journal
www.alpinejournal.org.uk

American Mountain Guide Association
www.amga.com

Ansel Adams Gallery
www.anseladams.com

Appalachian Mountain Club
www.outdoors.org

Appalachian Voices
www.appvoices.org

Aspen International Mountain Foundation
www.aimf.org

'Environnements et Montagnes: Savoirs et Politique' Research Group, University of Geneva
www.unige.ch/ses/geo/recherche/groupes/EquipeMontagnes

International Mountain Society
www.mrd-journal.org/ims.asp

Long Now Foundation
www.longnow.org

Messner Mountain Museum
www.messner-mountain-museum.it

Mountain Institute
www.mountain.org

Mountain Research Initiative
http://mri.scnatweb.ch/en

Mountain Societies Research Institute, University of Central Asia
http://msri.ucentralasia.org

Museo Nazionale della Montagna
www.museomontagna.org

Rocky Mountain Conservancy
www.rmconservancy.org

Royal Geographical Society
www.rgs.org

Sierra Club
www.sierraclub.org

UNESCO World Heritage Centre
www.whc.unesco.org

World Mountain People Association
www.mountainpeople.org

ACKNOWLEDGEMENTS

During the long trek that led to the completion of this small book, I have accumulated vast debts of gratitude towards a number of colleagues and friends. Special thanks go to Pete Adey and Felix Driver for offering their valuable feedback and suggestions on previous drafts of the book; to Father Maximos Constas, Bernard Debarbieux, Klaus Dodds, Ken Olwig, Al Pinkerton, Karen Pinto, Mario Pitteri and Lana Sloutsky for suggesting materials and ideas well beyond my horizons; to William Bainbridge and Stefano Cracolici for getting me acquainted with 'the Silver Age of mountaineering'; to Dan Webb for his precious research assistance; to Daniel Allen and Michael Leaman for their thorough editorial guidance and excellent suggestions, and to the Reaktion team, especially to Amy Salter and Becca Wright, for their great support during production. I am likewise extremely grateful to all the institutions and individuals who have granted their permission for the reproduction of images, and especially to my brother Luca for sharing his mountain photographs with me.

Drafts of some of the chapters were presented at various conferences, workshops and seminars. I am especially thankful to organizers Brita Staxrud Brenna and Torild Gjesvik at the University of Oslo, Betsey Robinson at Vanderbilt University, Christos Kakalis and Emily Goetsch at the University of Edinburgh, and to the participants of these events for their invaluable input. A modified version of Chapter Seven appeared in the volume *Moving Mountains: Studies in Performance, Changing Perspectives and Mobility*, edited by Christos Kakalis and Emily Goetsch (Basingstoke and New York, 2016).

Finally, I would like to thank the Fathers of Docheiariou Monastery, dwellers on the Holy Mountain, for their continuous support, and last but not least, my mountain-enthusiast mum for making me discover and appreciate the beauty of our Dolomites.

PERMISSION

For permission to reproduce the extract from 'Wise Men in their Bad Hours', p. 106, *The Collected Poetry of Robinson Jeffers*, vol. I: *1920–1928*, ed. Tim Hunt (Redwood City, CA, 1988). Permission granted by Stanford University Press.

PHOTO ACKNOWLEDGEMENTS

The author and publishers wish to express their thanks to the below sources of illustrative material and/or permission to reproduce it. Some permanent locations of works (or of temporary installations of works) are given below rather than in the captions.

Agence France Presse/Getty Images: p. 75; Alte Nationalgalerie, Berlin: p. 152; Alte Pinakothek, Munich (photo © Tarker/Bridgeman Images): p. 123; American School of Classical Studies at Athens (Gennadius Library): p. 192; photo Spencer Arnold, Getty Images: p. 185; © Ashmolean Museum, University of Oxford: pp. 158, 194, 198; photos author: pp. 16, 58, 99, 187, 208 (top); collection of the author: pp. 25, 96, 99; photo Jacques Beaulieu: p. 213; Beinecke Library, Yale University, New Haven, Connecticut: p. 126; Bibliotheek der Rijksuniversiteit, Leiden: p. 51; photo Maurus Blank: p. 208 (bottom); Bodleian Library, University of Oxford, Oxford: p. 20; British Library, London: pp. 52 (MS Or. 1525, ff. 8v-9r), 179; Brooklyn Museum, New York: pp. 161, 173; © Bundesarchiv/Bildarchiv, www.bild.bundesarchiv.de: pp. 103, 104; photo courtesy Simone Calò, www.simonecalo.com: p. 212 (bottom); Can Stock Photo Inc.: p. 113; from Sok-pil Cho, *There is No T'aebaeck Mountain Range* (Seoul, 1997): p. 22 (left); Collection Center for Creative Photography, University of Arizona, Tucson © 2015 The Ansel Adams Publishing Rights Trust: p. 206 (bottom); photos Luca della Dora: pp. 8, 24, 78, 226; Giuseppe di Tella, *Il bosco contro il torrente* (Milan, 1912): p. 202; photos Fr. Apollò of Docheiariou: pp. 19, 44; The Frick Collection, New York: p. 116; Groeningemuseum, Bruges, Belgium: p. 160; from J. B. Harley and David Woodward, *History of Cartography* (Chicago, 1987–) by permission of the University of Chicago Press: pp. 38, 39; Werner Harstad/The Norwegian Public Roads Administration and Norsk Form: pp. 216–17; Harvard University, Cambridge, Massachusetts (Houghton Library): pp. 66 (94-936), 68 (MTSD 0148); The Hereford Mappa Mundi Trust and the Dean and Chapter of Hereford Cathedral:

pp. 46, 47; photo courtesy Carl-Viggo Hølmebakk: p. 215; © Institut national de la propriété industrielle (INPI): p. 129; Istituto di Studi Bizantini e Post-bizantini, Venice: p. 65; drawing by Jung, 2003, adapted by Wonseob Song: p. 30; © karma_pema/iStockphoto LP: 95; Kenya Wildlife Service, Mount Kenya Official Guidebook (1890s?): p. 90; Kimbell Art Museum, Fort Worth, Texas: p. 62; photo Henning Klokkeråsen: p. 214; Korea University Museum, Seoul, South Korea: p. 22 (right); photo Alex Korting: pp. 211, 212 (top); Kunsthalle Hamburg, Germany: p. 101; © Look and Learn/Peter Jackson Collection, London: p. 132; photo © maxhomand/ 2016 iStock International, Inc.: p. 6; G. Mercator and J. Hondius, *Atlas* (Amsterdam, 1606) courtesy Geographicus Rare Antique Maps: p. 55; The Metropolitan Museum of Art, New York: pp. 37 (Fletcher Fund and Joseph E. Hotung and Michael and Danielle Rosenberg Gifts, 1989), 168, 172; Musée du Louvre, Paris: p. 80; Museo de la Casa Nacional de Moneda, Fundación Cultural BCB, Potosí, Bolivia: p. 48; Museum of Fine Arts, Boston: p. 206 (top); Museums Sheffield, Collection of the Guild of St George (Ruskin Collection), UK: p. 196 (bottom); NASA image by Robert Simmon, based on Landsat 5 data from the USGS Global Visualization Viewer: p. 200; NASA Earth Observatory: p. 209; NASA/MOLA Science Team: p. 15; National Archaeological Museum, Athens, Greece – photo Athanasios Meliarakis © Hellenic Ministry of Culture, Education and Religious Affairs/Archaeological Receipts Fund: p. 60; National Gallery, London: pp. 119, 143; National Palace Museum, Taipei, Taiwan: p. 135; New York Public Library, New York: pp. 153, 154; photo Finnbar O'Reilly/ Reuters: p. 94; photo Michael Olteanu, Christus Rex, Inc.: pp. 110–11; Philadelphia Museum of Art, Pennsylvania: p. 136; © Royal Geographical Society (with IBG), London: p. 89; Royal Museums of Fine Arts of Belgium, Brussels: p. 121; John Ruskin, *Modern Painters* (vol. IV): pp. 195 (p. 179), 196 (top) (p. 181); Johannes Schnitzer's illustration in Ptolemy's *Geographia* (Ulm, 1482): p. 21; photo Fr. Gaurav Shroff: p. 57; Smithsonian Institution, Washington, DC: p. 181; State Hermitage Museum, St Petersburg: p. 120; © State Tretyakov Gallery, Moscow: p.43; from Louise van Swaaij, Jean Klar and David Winner, *The Atlas of Experience* (London, 2000): p. 12; Swaggart Brothers Inc. www.swaggartbrothers.com: p. 140; Syndikat Verlag, Frankfurt: pp. 130, 131, 133; © Tate, London, 2016: p. 82 (top); The Temple Gallery, London: p. 134; photo A.B. Tilman: p. 69; Underwood & Underwood, United States Library of Congress's Prints and Photographs division: p. 190; © UNESCO: p. 207; United States Air Force, Washington, DC: p. 98; United States Geological Survey, Virginia: p. 148; University of Bristol Library (Special Collections): pp. 142, 150–51; © Victoria and Albert Museum, London: pp. 50, 117; © V&A Images/Alamy Stock Photo: p. 82 (bottom); The Walters Art Museum, Baltimore, Maryland: p. 26; photo Matt Wier: p. 220; and Zentralbibliothek Zürich, Switzerland (Map Department): pp. 127, 169.

code poet, the copyright holder of the photo on p. 70, Eric Hill, the copyright holder of the image on p. 180, and Tom Thai, the copyright holder of the photo on pp. 32–3, have published these online under conditions imposed by a Creative Commons Attribution-Share Alike 2.0 Generic license; M. Disdero, the copyright holder of the photo on p. 14, Joseph L. Hartman, the copyright holder of the photo on p. 77, and Eli Zahavi, the copyright holder of the image on p. 42, have published these online under conditions imposed by a Creative Commons Attribution-Share Alike 2.5 Generic license; Dennis Jarvis, the copyright holder of the photo on p. 76, Mountaineer, the copyright holder of the photo on p. 71, Nikmilano, the copyright holder of the photo on p. 97, Jean-Marc Rosier www.rosier.pro, the copyright holder of the photo on p. 114, Sailko, the copyright holder of the photo on p. 10, and Esben Stenfeldt, the copyright holder of the photo on p. 41, have published these online under conditions imposed by a Creative Commons Attribution-Share Alike 3.0 Unported license; Wellcome Library, London, the copyright holder of the image on p. 145, has published this online under conditions imposed by a Creative Commons Attribution 4.0 International license.

Readers are free:

to share – to copy, distribute and transmit the work(s)
to remix – to adapt the work(s)

under the following conditions:

you must attribute the work(s) in the manner specified by the author or licensor (but not in any way that suggests that they endorse you or your use of the work(s))

'share alike' – if you alter, transform, or build upon this work/these works, you may distribute the resulting work(s) only under the same or similar license to this one.

INDEX

Page numbers for illustrations are in *italics*